Contents

KV-499-943

MATHS NOW!
National Writing Group

JOHN MURRAY

£13.99
510
22445H

Acknowledgements

The authors and publishers would like to thank all the teachers, schools and advisers who evaluated *Maths Now! GCSE Intermediate 2* and whose comments contributed so much to this final version.

Particular thanks go to: David Bullock, Frodsham High School, Warrington; Philip Chaffé, King Edward VI School, Lichfield; John D. Collins, Education Consultant and Inspector of Schools; Patrick Gallagher, Convent of Jesus and Mary RC High School, London; Peter Marks, Ilfracombe College, Devon; Kevin Pankhurst, Pilton Community College, Barnstaple; Kim O'Driscoll-Tole, University of Strathclyde. Thanks are also due to Dr R.C. Solomon for the material on **pp. 16–17**.

Photo acknowledgements

pp. 10, 25, 40, 72, 85, 99, 100, 121, 145, 162, 209, 210, 212, 213, 220, 252, Screen shots reprinted by permission from Microsoft Corporation
p.1 *l* Mary Evans Picture Library, *r* Science Photo Library; **p.6** John Townson/Creation; **p.11** Library of Congress/Science Photo Library; **p.12** *l* The Art Archive/Album/ Joseph Martin, *r* Mary Evans Picture Library; **p.14** John Townson/Creation; **p.20** John Townson/Creation; **p.26** John Townson/Creation; **p.32** © Copyright Natwest; **p.57** *l* Mary Evans Picture Library, *r* © Copyright The British Library; **p.73** Mehaukulyk/ Science Photo Library; **p.87** *l* Mary Evans Picture Library, *r* Mark Winter (University of Sheffield and Webelements Ltd); **p.101** E. Hobson/Ancient Art; **p.123** *both* Mary Evans Picture Library; **p.146** *t* Mary Evans Picture Library, *c* Ancient Art, *b* Christian Andréason/ Robert Harding; **p.171** *l* Christopher Rennic/Robert Harding, *r* Robert Harding; **p.191** *both* © Copyright Cordon Art; **p.208** Rex Features Limited.; **p.242** Article and photo reproduced by kind permission of the Cambridge Evening News.

First published in 2002
by John Murray (Publishers) Ltd
50 Albemarle Street
London W1S 4BD

Layouts by Stephen Rowling/springworks
Artwork by Oxford Designers & Illustrators Ltd
Spreadsheet screen shots by R. Pimentel
Cover design by John Townson/Creation

Typeset in 10/12pt Times by Wearset Ltd, Boldon, Tyne and Wear
Printed and bound in Spain by Bookprint SL., Barcelona

A CIP catalogue record for this book is available from the British Library

ISBN 0 7195 7451 X
Maths Now! GCSE Intermediate 2 Teacher's Resource Book 0 7195 7452 8

Introduction

This textbook, and its companion volume *GCSE Mathematics Intermediate 1*, have been written by experienced classroom teachers to give you both enjoyment and a sense of achievement.

Each chapter has an introduction which tries to link mathematics to a wider context, be it historical, scientific, artistic or even dramatic.

The standard of the work has been deliberately set at a high level. This is for two reasons. The first is that those of you who wish to proceed with mathematics or science to a higher level will have studied the necessary mathematics to make this possible. The second reason is to make you think. To think hard.

Some questions may seem difficult at first. Try them before asking for help. Mathematics is about thinking logically. The cry from teachers to 'show your working' can be interpreted as a request to show the logical progress of your thinking.

We believe that this ability to solve problems by a process of logical thinking is a great asset in the modern world. You will need the process even if you do not need the detail.

The chapters in the book follow closely the order of your GCSE maths specification. However, it is unlikely that you will work through the book in page order. There are testing exercises at the end of each chapter, and supplementary work is also available. These are best done unaided.

We know that you will have studied mathematics for many years now. Some of the work in this book will be familiar but will extend your knowledge; other parts will be totally new. We hope that you will approach the work positively, expecting both difficulty and enjoyment.

Finally, we hope you will have a sense of achievement at the end of this course and that this achievement is reflected in the award of the highest grade at GCSE of which you are capable.

Symbols

The symbols used in the Student's Book are as follows:

 Use a calculator

 Use a graphics calculator

 Do not use a calculator

 Ma1

1 Indices and standard index form

Galileo Galilei 1564–1642

Galileo was an Italian astronomer and physicist. He was the first person accredited with having used a telescope to study the stars. in 1610 Galileo and a German astronomer, Marius, independently discovered Jupiter's four largest moons, Io, Europa, Ganymede and Callisto. At that time it was believed that the Sun revolved around the Earth. Galileo was one of the few people who believed that the Earth revolved around the Sun. As a result of this, the Church declared that he was a heretic and imprisoned him. It took the Church a further 350 years to accept officially that Galileo was correct; he was pardoned only in 1992.

Facts about Jupiter
Mass: 1 900 000 000 000 000 000 000 000 000 kg
Diameter: 142 800 000 m
Mean distance from the Sun: 779 000 000 km

Indices

The **index** refers to the power to which a number is raised. In 5^3 the number 5 is raised to the power of 3, which means $5 \times 5 \times 5$. The 3 is known as the index, while the 5 is known as the **base**. The plural of index is **indices**. Examples:

$$5^3 = 5 \times 5 \times 5 = 125$$
$$7^4 = 7 \times 7 \times 7 \times 7 = 2401$$
$$3^1 = 3$$

Remember:
$5^3 \leftarrow index$
$\quad \searrow base$

Laws of indices

When working with numbers involving indices there are three basic laws that can be applied. These are demonstrated below.

- $4^2 \times 4^4 = 4 \times 4 \times 4 \times 4 \times 4 \times 4$
  $= 4^6$ (i.e. 4^{2+4})

This can be written in a general form as:

$$a^m \times a^n = a^{m+n}$$

Note. The base numbers must be the same for this rule to be true.

- $3^6 \div 3^2 = \dfrac{3 \times 3 \times 3 \times 3 \times \cancel{3}^1 \times \cancel{3}^1}{\cancel{3}_1 \times \cancel{3}_1}$

  $= 3^4$ (i.e. 3^{6-2})

This can be written in a general form as:

$$a^m \div a^n = a^{m-n}$$

Note. The base numbers must be the same for this rule to be true.

- $(5^2)^3 = (5 \times 5) \times (5 \times 5) \times (5 \times 5)$
  $= 5^6$ (i.e. $5^{2 \times 3}$)

This can be written in a general form as:

$$(a^m)^n = a^{mn}$$

Examples   Simplify $4^3 \times 4^2$.

$4^3 \times 4^2 = 4^{(3+2)}$
$\qquad\quad = 4^5$

....

Simplify $(4^2)^3$.

$(4^2)^3 = 4^{(2 \times 3)}$
$\qquad\quad = 4^6$

....

Simplify $2 \times 2 \times 2 \times 5 \times 5$ using indices.

$2 \times 2 \times 2 \times 5 \times 5$
$= 2^3 \times 5^2$

Exercise 1.1

1 Using indices, *simplify* the following.
   a $3 \times 3 \times 3$
   b $6 \times 6 \times 6 \times 6$
   c $8 \times 8 \times 8 \times 8 \times 8 \times 8$
   d $4 \times 4 \times 4 \times 4 \times 4 \times 5 \times 5$
   e $3 \times 3 \times 3 \times 4 \times 4 \times 6 \times 6 \times 6 \times 6 \times 6$

2 Write out the following in full.
   a 4^2 **b** 5^7 **c** $4^3 \times 6^3$ **d** $7^2 \times 2^7$

> **Remember:**
> When **dividing** powers, **subtract** the indices.
> When **multiplying** powers, **add** the indices.

3 Simplify the following using indices.

a $3^2 \times 3^4$ **b** $8^5 \times 8^2$ **c** $4^3 \times 4^5 \times 4^2$ **d** $4^6 \div 4^2$

e $5^7 \div 5^4$ **f** $\dfrac{3^9}{3^2}$ **g** $(5^2)^2$ **h** $(3^5)^3$

4 Simplify the following.

a $\dfrac{2^2 \times 2^4}{2^3}$ **b** $\dfrac{3^4 \times 3^2}{3^5}$ **c** $f^5 \times f^3$ **d** $m^4 \div m^2$

e $(2p^2)^3$ **f** $(b^3)^5 \div b^6$

The zero index

The **zero index** means that a number has been raised to the power of 0. Any number raised to the power of 0 is equal to 1, that is:

$$4^0 = 1 \qquad 10^0 = 1 \qquad a^0 = 1$$

This can be explained by applying the laws of indices.

$$a^m \div a^n = a^{m-n}$$

Therefore

$$\frac{a^m}{a^m} = a^{m-m}$$

$$= a^0$$

However

$$\frac{a^m}{a^m} = 1$$

Therefore

$$a^0 = 1$$

> **Remember:**
> *Any number (except 0) divided by itself must be 1.*

> Any number to the power 0 is 1.

This can be shown using numbers in the following way:

$$5^3 = 5 \times 5 \times 5$$
$$5^2 = 5 \times 5$$
$$5^1 = 5$$
$$5^0 = 1$$

$\div 5$ $\div 5$ $\div 5$

Exercise 1.2

1 Without using a calculator, work out the value of each of the following.

a 2^5 **b** 3^4 **c** 8^2 **d** 6^3

e 10^6 **f** 4^4 **g** $2^3 \times 3^2$ **h** $10^3 \times 5^3$

2 Without using a calculator, evaluate the following.

a $2^3 \times 2^0$ **b** $5^2 \div 6^0$ **c** $(4^0)^2$ **d** $4^0 \div 2^2$

Exercise 1.3

1 Write an expression for the area of the square shown below.

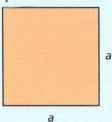

a

a

2 Write an expression for the total area of the shape shown below.

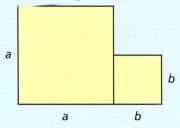

a

b

a   *b*

3 A cube of side length *c* units is shown below.

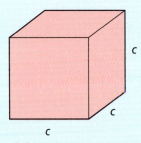

c

c

c

  a Write an expression for the volume of the cube.
  b Write an expression for the surface area of the cube.
  c What would be the value of *c* if the volume was $27 \, \text{cm}^3$?

4 A cuboid has the dimensions shown below.

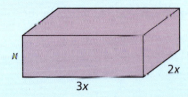

x

2x

3x

  a Write an expression for the volume of the cuboid.
  b Write an expression for the total surface area of the cuboid.
  c What is the value of *x* if the volume is $48 \, \text{cm}^3$?

Standard index form

If numbers are written in the normal way, they become increasingly more difficult to read and more laborious to write the larger they become.

This is demonstrated in the table below. It shows the mean distance from the Sun of each of the nine planets in our Solar System.

> **Remember:**
> *Mean is a type of average.*

planet	mean distance from Sun (km)
Mercury	58 000 000
Venus	108 000 000
Earth	149 000 000
Mars	228 000 000
Jupiter	778 000 000
Saturn	1 430 000 000
Uranus	2 870 000 000
Neptune	4 500 000 000
Pluto	5 910 000 000

To overcome this, a method of writing numbers in **standard index form** (often known as **standard form**) can be used. This involves writing numbers as a multiple of a power of 10. For example

$$100 = 1 \times 10^2$$
$$1000 = 1 \times 10^3$$
$$40\,000 = 4 \times 10\,000 = 4 \times 10^4$$

It is important to realise that $40\,000$ can be written in a number of different ways, for example

$$4 \times 10^4 \qquad 40 \times 10^3 \qquad 400 \times 10^2 \qquad 4000 \times 10^1 \quad \text{etc.}$$

> $\leqslant$ less than or equal to
> $<$ less than.

However, only 4×10^4 is written in standard index form. This is because, for a number to be in standard index form, it must take the form $A \times 10^n$, where n must be either a positive or negative integer (a whole number) and A must lie in the range $1 \leqslant A < 10$ (i.e. from 1 up to but not including 10).

Examples   Write 72 000 in standard form.

7.2×10^4

Of the numbers below, ring those which are written in standard form.

$\boxed{4.2 \times 10^3} \qquad 0.35 \times 10^2 \qquad 18 \times 10^5 \qquad \boxed{6 \times 10^3} \qquad 0.01 \times 10^1$

Multiply 600×4000 and write your answer in standard form.

600×4000
$= 2\,400\,000$
$= 2.4 \times 10^6$

Exercise 1.4

1 Which of the following are *not* in standard form?
  a 6.2×10^5 **b** 7.834×10^{16} **c** 8.0×10^5
  d 0.46×10^7 **e** 82.3×10^6 **f** 6.75×10^1

2 Write the following numbers in standard form.
  a 600 000 **b** 50 000 000 **c** 700 000 000 000 **d** 48 000 000
  e 784 000 000 000 **f** 534 000 **g** 7 million **h** 8.5 million

3 Write the following in standard form.
  a 68×10^5 **b** 720×10^6 **c** 8×10^5
  d 0.75×10^8 **e** 0.4×10^{10} **f** 50×10^6

4 Without using a calculator, multiply the following and give your answers in standard form.
  a 200×3000 **b** 6000×4000 **c** 7 million $\times 20$
  d 500×6 million **e** 3 million $\times 4$ million **f** $4500 \times 40 000$

Using a calculator

Check your calculator now!

Calculators are able to work with numbers entered in standard form. Often, with large numbers, they will present the answer in standard form too. An awareness of how a calculator operates is therefore important.

Different calculators have different keys for standard form. However, the two most common are Exp and EE .

Examples Multiply 2×10^5 and 4×10^3.

 = 800 000 000

In standard form this is written as 8×10^8.

Multiply 6×10^8 and 4×10^7.

 = 2.4×10^{16}

Note how your calculator displays an answer written in standard form.

Exercise 1.5

1 Using a calculator, work out the following and write your answers in standard form.

 a $(4.4 \times 10^3) \times (2 \times 10^5)$

 b $(6.8 \times 10^7) \times (3 \times 10^3)$

 c $(4 \times 10^5) \times (8.3 \times 10^5)$

 d $(5 \times 10^9) \times (8.4 \times 10^{12})$

 e $(8.5 \times 10^6) \times (6 \times 10^{15})$

 f $(5.0 \times 10^{12})^2$

> **Remember:**
> *Standard form must be written $A \times 10^n$ where $1 \leqslant A < 10$.*

2 Using a calculator, work out the following and write your answers in standard form.

 a $(3.8 \times 10^8) \div (1.9 \times 10^6)$

 b $(6.75 \times 10^9) \div (2.25 \times 10^4)$

 c $(6.3 \times 10^7) + (8.8 \times 10^5)$

 d $(3.15 \times 10^9) + (7.0 \times 10^6)$

 e $(5.3 \times 10^8) - (8 \times 10^7)$

 f $(6.5 \times 10^7) - (4.9 \times 10^6)$

3 Light from the Sun takes approximately 8 minutes to reach the Earth. If light travels at a speed of 3×10^8 m/s, calculate, to three significant figures, the distance from the Sun to the Earth.

4 Phileas Fogg in Jules Verne's book *Around the World in 80 Days* made a bet that he could, as the title suggests, travel around the world in 80 days. The circumference of the earth is 4×10^4 km. Calculate the average (i.e. mean) distance he would need to travel each day to win his bet.

The negative index

We have seen so far how standard index form can be used to write very large numbers. It can also be used to write very small numbers, as shown in the list below.

> **Remember:**
> $10^{-1} = \frac{1}{10}$
> $10^{-2} = \frac{1}{100}$
> $10^{-3} = \frac{1}{1000}$
> *etc.*

$$
\begin{aligned}
100 &= 1 \times 10^2 \\
10 &= 1 \times 10^1 \\
1 &= 1 \times 10^0 \\
0.1 &= 1 \times 10^{-1} \\
0.01 &= 1 \times 10^{-2} \\
0.001 &= 1 \times 10^{-3} \\
0.0001 &= 1 \times 10^{-4}
\end{aligned}
$$

> Remember the rule that a number in standard index form must take the form $A \times 10^n$ and that A must lie in the range 1 up to but not including 10.

Examples   Write 0.0032 in standard form.

3.2×10^{-3}

Write the following numbers in order of magnitude, starting with the largest.

   3.6×10^{-3} 5.2×10^{-5} 1×10^{-2} 8.35×10^{-2} 6.08×10^{-8}

For comparison it may be easier to write them out long hand.

$$
\begin{aligned}
3.6 \times 10^{-3} &= 0.0036 \\
5.2 \times 10^{-5} &= 0.000\,052 \\
1 \times 10^{-2} &= 0.01 \\
8.35 \times 10^{-2} &= 0.0835 \\
6.08 \times 10^{-8} &= 0.000\,000\,060\,8
\end{aligned}
$$

The order is therefore:

   8.35×10^{-2} 1×10^{-2} 3.6×10^{-3} 5.2×10^{-5} 6.08×10^{-8}

Exercise 1.6

1 Write down the following numbers in standard form.
   a 0.0006
   b 0.000053
   c 0.0000007
   d 0.0004145

2 Write down the following numbers in standard form.
   a 68×10^{-5}
   b 750×10^{-9}
   c 0.057×10^{-9}
   d 0.4×10^{-10}

3 Deduce the value of n in each of the following cases.
   a $0.00025 = 2.5 \times 10^n$
   b $0.00357 = 3.57 \times 10^n$
   c $0.0000006 = 6 \times 10^n$
   d $0.004^2 = 1.6 \times 10^n$

4 Write down the following numbers in order of size, starting with the largest.

$$3.2 \times 10^{-4}$$
$$6.8 \times 10^5$$
$$5.57 \times 10^{-9}$$
$$6.2 \times 10^3$$
$$5.8 \times 10^{-7}$$
$$6.741 \times 10^{-4}$$
$$8.414 \times 10^2$$

> **Remember:**
> The 10^n part gives the best indication of the magnitude (size) of the numbers.

5 Light travels at a speed of 3×10^8 m/s.
   a How long does it take for light to travel 1 metre?
   b Write your answer to part **a** in standard form.

SUMMARY

By the time you have completed this chapter, you should know:

■ what an **index** is

   in 5^4, the 4 is the power to which the **base** 5 is raised and is known as the index

■ the three basic laws of **indices**

   $a^m \times a^n = a^{m+n}$ $a^m \div a^n = a^{m-n}$ $(a^m)^n = a^{mn}$

■ that any number to the power of zero is 1

   $a^0 = 1$

■ why **standard form (standard index form)** is used and why it is useful
■ how to write numbers in standard form

   $A \times 10^n$ where A lies in the range from 1 up to but not including 10, and n is a whole number

■ how to carry out calculations in standard form using a calculator, i.e. using the keys Exp or EE
■ how to write numbers between 0 and 1 in standard form

   in the general rule $A \times 10^n$, n is negative

Exercise 1A

1 Using indices, simplify the following.
   a $2 \times 2 \times 2 \times 5 \times 5$ **b** $2 \times 2 \times 3 \times 3 \times 3 \times 3 \times 3$
2 Write the following in full.
   a 4^3 **b** 6^4
3 Work out the value of the following without using a calculator.

   a $2^3 \times 10^2$ **b** $1^4 \times 3^3$ **c** $\dfrac{3^5}{3^3}$

4 Simplify the following using indices.

   a $3^4 \times 3^3$ **b** $6^3 \times 6^2 \times 3^4 \times 3^5$ **c** $\dfrac{4^5}{2^3}$ **d** $\dfrac{(6^2)^3}{6^5}$

5 Write the following numbers in standard form.
   a 8 million   **b** $0.000\,72$
6 Write the following numbers in order of magnitude, starting with the smallest.

      6.2×10^7 5.5×10^{-3} 4.21×10^7 4.9×10^8 3.6×10^{-5}

7 The speed of light is 3×10^8 m/s. Venus is 108 million km from the Sun. Calculate the number of minutes it takes sunlight to reach Venus.
8 A star system is 500 light years away from Earth. If the speed of light is 3×10^5 km/s, calculate the distance of the star system from Earth. Give your answer in kilometres and written in standard form. (**Note.** A light year is how far light travels in 1 year.)

Exercise 1B

1 Using indices simplify the following.
   a $3 \times 2 \times 2 \times 3 \times 27$ **b** $2 \times 2 \times 4 \times 4 \times 4 \times 2 \times 32$
2 Write the following out in full.
   a 6^5 **b** 9^4
3 Work out the value of the following without using a calculator.

   a $3^3 \times 10^3$ **b** $1^5 \times 4^3$ **c** $\dfrac{4^6}{4^4}$

4 Simplify the following using indices.

   a $2^4 \times 2^3$ **b** $7^5 \times 7^2 \times 3^4 \times 3^8$ **c** $\dfrac{4^8}{2^{10}}$ **d** $\dfrac{(3^3)^4}{27^3}$

5 Write the following numbers in standard form.
   a 460 million   **b** 3
6 Deduce the value of n in each of the following.
   a $4750 = 4.75 \times 10^n$ **b** $0.0040 = 4.0 \times 10^n$
7 The speed of light is 3×10^8 m/s. Jupiter is 778 million km from the Sun. Calculate the number of minutes it takes sunlight to reach Jupiter.
8 A star system is 300 light years away from Earth. If the speed of light is 3×10^5 km/s, calculate the distance of the star system from Earth. Give your answer in kilometres and written in standard form.

Exercise 1C

Here is a sequence of square patterns.

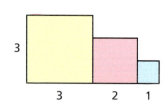

a Work out the total area of each of the three patterns.
b Draw the next three patterns in the sequence and calculate their total areas.
c Copy and complete the table below for the first six patterns of the sequence.

pattern	1	2	3	4	5	6
total area						

d Explain how the terms of your sequence are changing.
e Predict the total area of the 10th pattern in this sequence.

Exercise 1D

In exercise 1C you investigated a sequence of areas generated by a square pattern. Use a spreadsheet to generate the results you obtained and use formulae where appropriate.

Your spreadsheet may look similar to the one shown below:

You will need:
• computer with spreadsheet package installed

	A Pattern	B Area of new square	C Area of total shape
2	1	1	1
3	2	4	5
4	3	9	
5	4		
6	5	Use a formula to work out this column	Use a formula to work out this column
7	6		
8	7		
9	8		
10	9		
11	10		

Extend your spreadsheet to work out at which pattern the total area will be 11 440 units^2.

Exercise 1E

Using the internet as a resource, find out what theory Galileo and Copernicus shared.

You will need:
• computer with internet access

Nicolaus Copernicus 1473–1543

2 Solving numerical problems

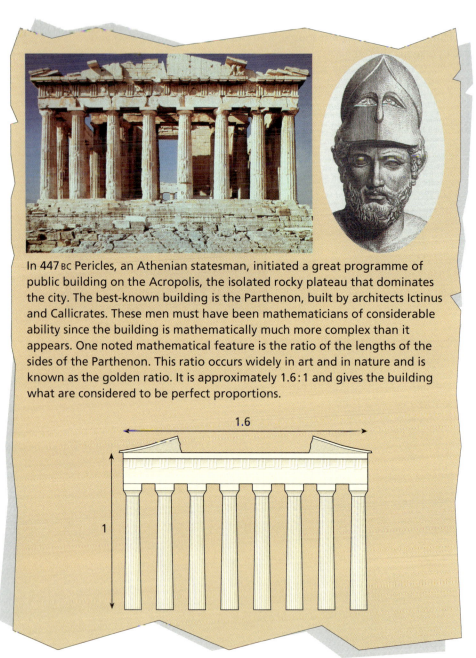

In 447 BC Pericles, an Athenian statesman, initiated a great programme of public building on the Acropolis, the isolated rocky plateau that dominates the city. The best-known building is the Parthenon, built by architects Ictinus and Callicrates. These men must have been mathematicians of considerable ability since the building is mathematically much more complex than it appears. One noted mathematical feature is the ratio of the lengths of the sides of the Parthenon. This ratio occurs widely in art and in nature and is known as the golden ratio. It is approximately 1.6 : 1 and gives the building what are considered to be perfect proportions.

Remember:
Where numbers repeat in a recurring decimal, place a dot over the first and last numbers of the recurring sequence; for example 7.128 128 128 . . . is written as 7.1̇28̇.

In *Intermediate 1* you learned how to:

- convert **terminating** and **recurring decimals** to fractions
- calculate the percentage of a quantity
- express one quantity as a percentage of another
- carry out calculations involving percentage increase and decrease
- solve problems involving direct proportion
- divide a quantity in a given ratio.

Exercise 2.1

(*Revision*)

1 Convert each of the following decimals to a fraction in its lowest terms.
   a 0.75   **b** 0.875
2 Convert the following fractions to recurring decimals.
   a $\frac{2}{3}$ **b** $\frac{13}{99}$
3 Work out the following.
   a 17% of 50   **b** 35% of 280
4 Express each of the following as a percentage.
   a 24 out of 50   **b** $\frac{13}{20}$
5 Increase £3000 by 8%.
6 Decrease 130 by 15%.
7 Tin and copper in an alloy are mixed in the ratio 8:3. How much tin is needed to mix with 36 g of copper?
8 Divide 20 m in the ratio 3:2.

> **Remember:**
> *17% can be written as 0.17.*

Exercise 2.2

(*Revision*)

1 Convert each of the following to a percentage.
   a 1.85   **b** $\frac{4}{5}$
2 Convert the following fractions to recurring decimals.
   a $\frac{71}{99}$ **b** $4\frac{1}{3}$
3 Calculate 75% of 360.
4 Express 420 as a percentage of 280.
5 Increase 20 by 20%.
6 Decrease 400 by 17.5%.
7 Divide 54 in the ratio 5:3:1.
8 Divide 390 in the ratio $\frac{1}{2}:\frac{1}{3}:\frac{1}{4}$.

Reciprocals

The **reciprocal** of a number is 1 divided by that number. So, the reciprocal of 7 is $\frac{1}{7}$.

● *As a general rule, where x is not zero:*

$$\text{the reciprocal of } x \text{ is } \frac{1}{x} \text{ or, in index notation, } x^{-1}$$

This is true for whole numbers and fractions.

Examples   Find the reciprocal of $\frac{3}{4}$.

By definition, the reciprocal is $1 \div \frac{3}{4}$.
$\qquad 1 \div \frac{3}{4}$ can be written as $1 \times \frac{4}{3}$.
Therefore $1 \div \frac{3}{4} = \frac{4}{3}$.

> When finding the reciprocal of a fraction, a quick way of calculating the answer is to swap the numerator and the denominator around.

Find the reciprocal of $3\frac{4}{5}$.

Firstly convert to an improper fraction:
$\qquad 3\frac{4}{5} = \frac{19}{5}$
Therefore the reciprocal of $\frac{19}{5} = \frac{5}{19}$.

Exercise 2.3

Find the reciprocal of each of the following.

1 5

2 12

3 $\frac{1}{9}$

4 $\frac{4}{5}$

5 100

6 $2\frac{1}{2}$

7 0.25

8 1.6

9 y

> **Remember:**
> If $x \neq 0$ then x^{-1} or $\frac{1}{x}$ is the reciprocal of x.

10 $3p$

11 $\dfrac{2m}{5}$

12 $\dfrac{r}{s}$

Using a calculator to find reciprocals

All scientific calculators have a reciprocal key. It will look like $\boxed{x^{-1}}$ or $\boxed{^1/_x}$.

Example Use a calculator to work out the reciprocal of $\frac{2}{7}$.

$$\boxed{2}\ \boxed{a^b/_c}\ \boxed{7}\ \boxed{x^{-1}}\ \boxed{=}\ 3.5$$

Note that most calculators will give the solution as a decimal.

Exercise 2.4

1 Boxes A and B below contain numbers. The numbers in box B are reciprocals of those in box A.

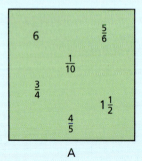

A

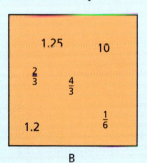

B

a Pair each of the numbers in box A to its reciprocal in box B.
b Using a calculator if necessary, multiply each pair of numbers together.
c What do you notice about the product of a number and its reciprocal?
d If the numbers in box B are the reciprocal of those in box A, are the numbers in box A the reciprocal of those in box B? Explain your answer carefully.

2 Work out the value of the letter in each of the following.

a $\frac{2}{3} \times a = 1$ **b** $b \times 5 = 1$

c $\frac{3}{8} \times \frac{8}{3} = c$ **d** $(7 \times \frac{1}{7}) - (\frac{4}{9} \times \frac{9}{4}) = d$

Rational and irrational numbers

The student who 'invented' irrational numbers was drowned in the Aegean Sea by his Greek tutors for spoiling their theories of completeness.

A **rational number** is any number which can be expressed as a fraction. Examples of some rational numbers and how they can be expressed as a fraction are shown below:

$$0.2 = \frac{2}{10} = \frac{1}{5} \quad 0.3 = \frac{3}{10} \quad 7 = \frac{7}{1} \quad 1.53 = \frac{153}{100} \quad 0.\dot{2} = \frac{2}{9}$$

An **irrational number** cannot be expressed as a fraction. Examples are:

$$\sqrt{2} \quad \sqrt{5} \quad 6 - \sqrt{3} \quad \pi$$

Rational numbers therefore include:

- integers
- fractions
- recurring decimals
- terminating decimals.

Irrational numbers include:

- the square root of any number other than a square number
- a decimal number which neither repeats nor terminates, for example π.

Exercise 2.5

1 Explain in your own words why the square root of a square number is rational.

2 Explain in your own words why π is classified as an irrational number.

3 For each of the numbers below, state whether it is rational or irrational, and give reasons why.

a 1.3 **b** $0.\dot{6}$ **c** $\sqrt{3}$ **d** $-2\frac{3}{5}$ **e** $\sqrt{25}$

f $\sqrt[3]{8}$ **g** $\sqrt{7}$ **h** 0.625 **i** $0.\dot{1}\dot{1}$

4 For each of the numbers below, state whether it is rational or irrational, and give reasons why.

a $\sqrt{2} \times \sqrt{3}$ **b** $\sqrt{2} + \sqrt{3}$ **c** $(\sqrt{2} \times \sqrt{3})^2$ **d** $\dfrac{\sqrt{8}}{\sqrt{2}}$

5 In each of the following, decide whether the value required is rational or irrational. Give reasons for your answer.

a The circumference of the circle   **b** The length of the diagonal   **c** The area of the triangle

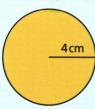

4 cm

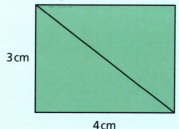

3 cm

4 cm

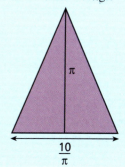

π

$\dfrac{10}{\pi}$

Surds

A **surd** is any expression involving roots of rational numbers. Examples of surds are

$$2 + \sqrt{3} \qquad 7 - \sqrt{2} \qquad 2\sqrt{5} + 4\sqrt{7} \qquad \tfrac{1}{2}\sqrt{\tfrac{3}{4}}$$

When surds are combined by arithmetical operations, the results can sometimes be simplified. Note that for example:

$$\sqrt{2} \times \sqrt{2} = 2 \qquad \sqrt{2} \times \sqrt{3} = \sqrt{6} \qquad 3\sqrt{2} + 5\sqrt{2} = 8\sqrt{2}$$

Note. When we multiply the square roots of two numbers, the result is the square root of the product.

$$\sqrt{a} \times \sqrt{b} = \sqrt{ab}$$

There is no way to simplify the *sum* of two square roots.

$$\sqrt{a} + \sqrt{b} \neq \sqrt{a + b}$$

Just try any pair of non-zero numbers. For example $\sqrt{2} + \sqrt{3} \neq \sqrt{5}$.

Examples Simplify $\sqrt{3} + \sqrt{12}$.

Note that

$$\begin{aligned} \sqrt{12} &= \sqrt{(4 \times 3)} \\ &= \sqrt{4} \times \sqrt{3} \\ &= 2 \times \sqrt{3} \end{aligned}$$

Hence $\sqrt{3} + \sqrt{12} = \sqrt{3} + 2\sqrt{3}$.
Therefore $\sqrt{3} + \sqrt{12} = 3\sqrt{3}$.

....

Expand and simplify $(2 + \sqrt{3})(3 - \sqrt{3})$.

Multiply both terms in the first bracket by both terms in the second bracket.

$$(2 + \sqrt{3})(3 - \sqrt{3}) = 2 \times 3 - 2 \times \sqrt{3} + 3 \times \sqrt{3} - \sqrt{3} \times \sqrt{3}$$

Note that $-2 \times \sqrt{3} + 3 \times \sqrt{3} = 1 \times \sqrt{3}$ and that $\sqrt{3} \times \sqrt{3} = 3$.

$$(2 + \sqrt{3})(3 - \sqrt{3}) = 6 + 1 \times \sqrt{3} - 3$$

Hence $(2 + \sqrt{3})(3 - \sqrt{3}) = 3 + \sqrt{3}$.

Exercise 2.6

1 Simplify these expressions.

  a $\sqrt{3} \times \sqrt{3}$ **b** $\sqrt{5} \times \sqrt{5}$ **c** $\sqrt{3} + \sqrt{3}$

  d $\sqrt{2} + \sqrt{2}$ **e** $3\sqrt{5} - \sqrt{5}$ **f** $4\sqrt{7} + 3\sqrt{7}$

2 Simplify these expressions.

  a $\sqrt{2} + \sqrt{8}$ **b** $\sqrt{7} + \sqrt{63}$ **c** $\sqrt{20} + \sqrt{45}$

  d $3\sqrt{2} - 4\sqrt{8}$ **e** $5\sqrt{10} - \sqrt{40}$ **f** $\sqrt{28} - \sqrt{7}$

3 Expand these expressions and simplify as far as possible.

  a $(3 + \sqrt{2})(1 + \sqrt{2})$ **b** $(2 - \sqrt{2})(3 + \sqrt{2})$ **c** $(5 + \sqrt{5})(3 - \sqrt{5})$

  d $(1 + 2\sqrt{3})(4 - 3\sqrt{3})$ **e** $(3 + 3\sqrt{2})(5 - 2\sqrt{2})$ **f** $(3 - 2\sqrt{5})(4 - 3\sqrt{5})$

Surds arise in many mathematical contexts, in particular when using Pythagoras' theorem or in trigonometry. It is often more appropriate to leave a result as a surd rather than to evaluate it to a certain number of significant figures (unless of course you are asked to give the answer to a certain number of significant figures).

Example Find the unknown length x in the diagram shown.

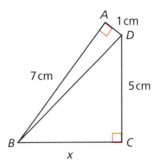

First find the hypotenuse BD of the top triangle. By Pythagoras' theorem, it is equal to

$$\sqrt{7^2 + 1^2} = \sqrt{50}$$

Leave this as $\sqrt{50}$. Apply Pythagoras' theorem in the right-hand triangle.

$$x = \sqrt{(\sqrt{50})^2 - 5^2} = \sqrt{50 - 25} = \sqrt{25} = 5$$

The unknown side has length exactly 5 cm.

Note. If we had evaluated $\sqrt{50}$ as 7.07, and then squared this, we would have obtained

$$\sqrt{7.07^2 - 5^2} = \sqrt{49.98 - 25} = \sqrt{24.98} = 4.998$$

Though this is very close to 5, it is not exactly so. The final answer should be exactly 5 cm.

Exercise 2.7

Find the unknown lengths in these diagrams. Where appropriate, leave your answer as a surd.

1

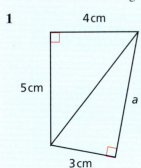

2

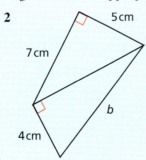

3

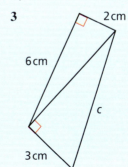

> **Remember:**
> *To change a percentage to a decimal, divide by 100.*

Percentage increase and decrease

In *Intermediate 1* you learned that calculations involving percentage increases and decreases can be carried out using a single multiplier.

Examples

A woman has a salary increase of 12%. If her original salary was £24 000 per year, calculate her new salary.

Her present salary of £24 000 represents 100%.
Her new salary will therefore be equivalent to 100% plus the 12% increase (i.e. 112% of her original salary).
112% is equivalent to a multiplier of ×1.12.
Therefore her new salary is £24 000 × 1.12 = £26 880.

The price of a new car is discounted by 5%. If its original price was £14 000, calculate the discounted price.

The original price of £14 000 represents 100%.
The discounted price is therefore equivalent to 100% minus the 5% discount (i.e. 95% of the original price).
95% is equivalent to a multiplier of ×0.95.
Therefore the discounted price is £14 000 × 0.95 = £13 300.

Exercise 2.8

1 A woman decides to invest in some shares. She buys 1000 shares in a company at a cost of £3.50 each. At the end of the year the price of shares in the company has increased in value by 15%.
   a Calculate the initial cost of buying the 1000 shares.
   b What is the total value of the shares at the end of the year?
   c If she decides to sell her shares at the end of the year, how much profit has she made?
2 Because of the flooding that occurred in the UK in November 2000, the value of many houses dropped.
   a If a house was originally worth £65 000, calculate what it was worth after the flooding, if its value decreased by 45%.
   b How much has the value of the house decreased (in £)?
3 A farmer owns 50 km² of land. Initially 30 km² of his land is used for growing wheat. He then decides to increase the percentage of his land used for wheat by 12%.
   a What percentage of his land was originally used for growing wheat?
   b How many km² of land are used for growing wheat after the increase?
4 Carry's and Doxon's are two high street computer stores. Carry's are selling a computer for £720. Doxon's are selling the same computer for £690. During a sale, the stores have the following signs on their windows:
   a Calculate the price of the computer in Carry's sale.
   b Calculate the price of the computer in Doxon's sale.

Carry's Sale
Everything
35% OFF

Doxon's Sale

Everything
30% OFF

Reverse percentages

A percentage increase or decrease is always given as a percentage of the original quantity. Therefore if we are told the new quantity after a percentage increase or decrease, it is possible to work backwards and calculate the original amount. This type of calculation is known as a **reverse percentage** calculation, as it goes from the new amount back to the original.

Examples   A three-year-old car was sold for a trade-in price that was 60% of its price when new. If its trade-in price was £7200, what did it cost when new?

> 60% is £7200
> 1% is $\frac{7200}{60}$
> Therefore 100% is $\frac{7200}{60} \times 100$

The cost of the car when new was £12 000.

In a test, Ahmed answered 92% of the questions correctly. If he answered 23 questions correctly, how many had he got wrong?

> 92% is 23 questions
> 1% is $\frac{23}{92}$
> 100% is $\frac{23}{92} \times 100 = 25$ questions

Therefore Ahmed got two questions wrong.

A boat is sold for £15 360. This represents a profit of 28% to the seller. What did the boat originally cost the seller?

> 128% of the original cost is £15 360
> 1% is therefore $\frac{15360}{128}$
> 100% is therefore $\frac{15360}{128} \times 100 = 12000$

So the boat was £12 000 when bought originally.

Exercise 2.9

1 Calculate the value of X in each of the following.
   a 40% of X is 240   **b** 24% of X is 84
   c 15% of X is 18.75   **d** 7% of X is 0.105

2 Calculate the value of Y in each of the following.
   a 125% of Y is 70   **b** 140% of Y is 91
   c 150% of Y is 0.375 **d** 144% of Y is −54.72

> **Remember:**
> *Reverse percentage goes from new to old, instead of from old to new.*

3 In a geography textbook, 35% of the pages are coloured. If there are 98 coloured pages, how many pages are there in the whole book?

4 A town has 3500 families who own a car. If this represents 84% of the families in the town, how many families are there in total?

5 In a test, Isabel scored 88%. She got three questions wrong. How many questions did Isabel get right if all questions carried equal marks?

6 Water expands when it freezes. Ice is less dense than water so it floats. If the increase in volume is 4%, what volume of water will make an iceberg of volume 12 700 000 m³? Give your answer correct to three significant figures.

Ratio and proportion

You know from *Intermediate 1* that ratios operate in a similar way to fractions.

In the diagram above, the strip is split into two colours.

The yellow strip is $\frac{3}{4}$ the size of the red strip.
The red strip is $\frac{4}{3}$ the size of the yellow strip.
The yellow strip is also $\frac{3}{7}$ of the total.
The red strip is $\frac{4}{7}$ of the total.

> **Remember:**
> *Numbers written in ratios must always be in the same units.*

All of these fractions can be deduced from the fact that the strip is divided in the ratio 3:4.

Being aware of the relationship between ratio and fractions will help you carry out calculations involving ratio.

Examples

A photograph is 12 cm wide and 8 cm tall. It is enlarged in the ratio 3:2. What are the dimensions of the enlarged photograph?

3:2 is an enlargement of $\frac{3}{2}$. Therefore each dimension on the original photograph is multiplied by $\frac{3}{2}$.
Enlarged width is $\frac{3}{2} \times 12 = 18$ cm.
Enlarged height is $\frac{3}{2} \times 8 = 12$ cm.

A photographic transparency 5 cm and 3 cm tall is projected onto a screen. The projected image is 1.5 m wide.
a Calculate the ratio of enlargement.
b Calculate the height of the image.

a 5 cm width is enlarged to become 150 cm.
The enlargement factor is therefore $\frac{150}{5}$, i.e. $\times 30$.
The enlargement ratio is therefore 30:1.
b The height of the image is $3 \times 30 = 90$ cm.

Exercise 2.10

1 Increase 40 by a ratio of 5:4.
2 Decrease 40 by a ratio of 4:5.
3 Increase 150 by a ratio of 7:5.
4 Decrease 210 by a ratio of 3:7.
5 A photograph 8 cm by 6 cm is enlarged in the ratio 11:4. What are the dimensions of the enlarged print?
6 A photocopier enlarges in the ratio 7:4. What is the enlarged size of a diagram that was originally 16 cm by 12 cm?
7 A drawing measuring 10 cm by 16 cm needs to be enlarged to 25 cm by 40 cm. Calculate the factor of enlargement needed and express it as a ratio.
8 A banner needs to be enlarged from the original design. The banner in the original design is 4 cm tall by 25 cm wide. The enlarged banner needs to be at least 8 m wide, but no more than 1.4 m tall. Calculate the minimum and maximum possible ratios of enlargement.

Repeated proportional change

When a proportional change, such as an enlargement or reduction, is repeated several times, calculations can become quite laborious. However, with an understanding of indices the process is much more efficient.

For example, deposit accounts in banks usually pay out interest. In other words, if money is left in the account for a period of time, the amount of money in the account increases.

In this example, the savings account offers 6% interest each year.

NOTWEST BANK
Super-Saver Account
Interest:
6% per annum

If a customer deposits £1000, then after one year this will have grown to £1000 × 1.06 = £1060.

If we are asked to work out the amount after five years, then the calculation becomes lengthy, i.e.

£1000 × 1.06 × 1.06 × 1.06 × 1.06 × 1.06 = £1338.23

However, if we use indices, the calculation becomes

£1000 × 1.06^5 = £1338.23

Example   A picture measures 26 cm by 18 cm. It can be repeatedly enlarged in the ratio 5:4. How many enlargements are necessary so that the enlarged picture best fits a frame 52 cm by 36 cm?

First enlargement is $26 \times \frac{5}{4} = 32.5$ cm
by
$18 \times \frac{5}{4} = 22.5$ cm

Second enlargement is $26 \times (\frac{5}{4})^2 = 40.6$ cm
by
$18 \times (\frac{5}{4})^2 = 28.1$ cm

Third enlargement is $26 \times (\frac{5}{4})^3 = 50.8$ cm
by
$18 \times (\frac{5}{4})^3 = 35.2$ cm

Fourth enlargement is $26 \times (\frac{5}{4})^4 = 63.5$ cm
by   → Too big
$18 \times (\frac{5}{4})^4 = 43.9$ cm

Therefore three enlargements are necessary for the picture to fit the frame best.

Exercise 2.11

1 A map measures 15 cm by 10 cm. It is enlarged four times in the ratio 5:4. Calculate the final dimensions of the map.
2 The population in a city is projected to grow at a steady rate of 8% per year. If its population is currently 650 000, calculate the city's population in five years' time. Give your answer to three significant figures.
3 A bacterial infection grows at a rate of 50% per day. It currently occupies an area of 5 cm².
   a Estimate the area it will cover if left untreated for 8 days.
   The bacteria can be treated using antibiotics. The effect of the antibiotics is to reduce the number of bacteria by 80% each day.
   b Assuming the course of antibiotics is started on the eighth day, calculate how long it will take for the bacteria to cover an area of approximately 1 cm².
4 A photocopier enlarges in a ratio of 7:4. A picture measures 6 cm by 4 cm. How many consecutive enlargements need to be made so that the enlarged picture will best fit a frame 30 cm by 20 cm?
5 A cube is enlarged in the ratio of 3:2.

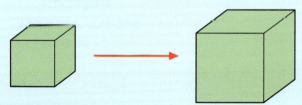

   a Write down the ratio Volume of enlarged cube : Volume of original cube in its simplest form.
   b Write down the ratio Surface area of enlarged cube : Surface area of original cube in its simplest form.

SUMMARY

By the time you have completed this chapter, you should know:

■ how to find the **reciprocal** of a number; for example

reciprocal of 5 is $\frac{1}{5}$,   reciprocal of $\frac{4}{7}$ is $\frac{7}{4}$

■ what is meant by **rational numbers**, **irrational numbers** and **surds**
■ how to increase and decrease a quantity by a percentage; for example

increase £150 by 20%,   decrease £200 by 10%

■ how to use **reverse percentages** to calculate a starting value; for example

A house has decreased in value by 15% since last year. If it is now worth £75 000, what was its value last year?

■ how to increase and decrease in a given ratio; for example

a photograph measuring 15 cm by 10 cm is enlarged in the ratio 3 : 2

■ how to calculate **repeated proportional change**; for example

£1000 was deposited in a savings account with an interest rate of 5% per year. How much will it be worth after eight years?

Exercise 2A

1 Calculate the selling price for each of the following items, given the original price and the profit made.

	original price	profit
a	£3500	8%
b	£12 000	24%
c	£1	250%

2 Calculate the original price in each of the following.

	selling price	loss
a	£350	30%
b	£200	20%
c	£8000	60%

3 In a test, Ben gained 90% by answering 135 questions correctly. If all questions carried equal marks, how many did he answer incorrectly?
4 A photograph 8 cm by 6 cm is enlarged in the ratio of 11 : 4. What are the dimensions of the enlarged print?
5 A photocopier enlarges in the ratio 5 : 4. What would be the enlarged size of a diagram that was originally 16 cm by 12 cm?
6 A photograph measuring 5 cm by 8 cm needs to be enlarged to 25 cm by 40 cm. Calculate the enlargement needed and express it as a ratio.

Exercise 2B

1 A one year-old car is worth £11 250. If its value had decreased by 25% in that first year, calculate its price when new.
2 This year a farmer's crop yielded 50 000 tonnes. If this represents a 20% increase on last year, what was his yield last year?
3 A company increased its productivity by 10% each year for the last two years. If it produced 56 265 units this year, how many units did it produce two years ago?
4 A banner needs to be enlarged from its original design. The banner in the original design is 8 cm tall by 20 cm wide. The enlarged banner needs to be at most 5 m wide but no more than 1.8 m tall. Calculate the maximum possible ratio of enlargement.
5 A rectangle 7 cm by 4 cm is enlarged by a ratio of 2 : 1.
  a What is the area of:
    i) the original rectangle
    ii) the enlarged rectangle?
  b What is the ratio of the area of the enlarged rectangle to the area of the original rectangle?
6 A map measuring 60 cm by 25 cm is reduced twice in the ratio 3 : 5. Calculate the final dimensions of the map.

Exercise 2C

You will need:
• squared paper

The chequered flag below is 7 units by 5 units.

a How many squares are black?
b How many squares are white?
c Investigate similar chequered flags to find how many squares are black or white on a flag p units by q units.

Exercise 2D

The fractions below form a sequence:

$$1 \quad \frac{1}{2} \quad \frac{1}{4} \quad \frac{1}{8} \quad \frac{1}{16}$$

a Describe how the sequence is formed.
b Write down the next two terms of the sequence.
c Construct a spreadsheet, with a formula, to generate this sequence of fractions.
d Enter a formula to find the sum of this sequence. For example, your spreadsheet after the first four terms might look like this:

You will need:
- computer with spreadsheet package installed

Position	Term	Sum
1	1	1
2	1/2	1 1/2
3	1/4	1 3/4
4	1/8	1 7/8

e What happens to the sum of the sequence as the number of terms increases?

Exercise 2E

Using the internet as a resource, find examples of where the golden ratio appears naturally in nature.

You will need:
- computer with internet access

3 Time and money

The system of time keeping we use today is extremely precise. It is based on our movement around the Sun, and it is so accurate that it will take 44 000 years before it falls out of step with the Sun by as much as a single day.

Although all countries use a common unit for measuring the passage of time, we do not all use the same reference point (i.e. year 0). Christians use the birth of Christ as their reference point. So, in the Christian world, we are in the third millennium. Orthodox Jews take their reference point as the day that they believe the Earth was created, i.e. 6 October 3761 BC. Muslims, on the other hand, calculate the date from the year after Mohammed's flight to Mecca in AD 622, while Hindus calculate from the birth of Bramha.

The 24-hour clock

Most everyday times are given in terms of the 12-hour clock. We tend to say things like 'I got up at seven o'clock this morning, I went to play football at two o'clock in the afternoon and I went to bed at eleven o'clock.' These times can be written as 7 a.m., 2.00 p.m. and 11 p.m.

Uckfield, East Grinstead and Oxted to London ⑫

London, East Croydon, Oxted, East Grinstead and Uckfield

Sunday

		SN	SN	SN	SN	SN	SN	SN	SN	SN	SN	SN	SN	SN	SN	SN	SN	SN
London Victoria	dep	....	0726	0826	0926	1026	1126	1226	1326	1426	1526	1626	1726	1826	1926	2026	2126	2236
Clapham Junction	dep	....	0732u	0832u	0932u	1032u	1132u	1232u	1332u	1432u	1532u	1632u	1732u	1832u	1932u	2032u	2132u	2242u
London Bridge	dep	....	0710b	0813b	0925b	1025b	1125b	1225b	1325b	1425b	1525b	1625b	1725b	1825b	1925b	2025b	2125b	2225b
Norwood Junction	dep	....	0748	0848	0948	1048	1148	1248	1348	1448	1548	1648	1748	1848	1948	2048	2148	2259
East Croydon	dep	....	0753	0853	0953	1053	1153	1253	1353	1453	1553	1653	1753	1853	1953	2053	2153	2305
South Croydon	dep	....																
Sanderstead	dep	....	0757	0857	0957	1057	1157	1257	1357	1457	1557	1657	1757	1857	1957	2057	2157	2309
Riddlesdown	dep	....	0759	0859	0959	1059	1159	1259	1359	1459	1559	1659	1759	1859	1959	2060	2159	2311
Upper Warlingham	dep	....	0803	0903	1003	1103	1203	1303	1403	1503	1603	1703	1803	1903	2003	2103	2203	2315
Woldingham	dep	....	0807	0907	1007	1107	1207	1307	1407	1507	1607	1707	1807	1907	2007	2107	2207	2319
Oxted	arr	....	0812	0912	1012	1112	1212	1312	1412	1512	1612	1712	1812	1912	2012	2112	2212	2324
Oxted	dep	0002	0812	0912	1012	1112	1212	1312	1412	1512	1612	1712	1812	1912	2012	2112	2212	2324
Hurst Green	dep	0004	0814	0914	1014	1114	1214	1314	1414	1514	1614	1714	1814	1914	2014	2114	2214	2326
Lingfield	dep	0010	0820	0920	1020	1120	1220	1320	1420	1520	1620	1720	1820	1920	2020	2120	2220	2332
Dormans	dep	0013	0823	0923	1023	1123	1223	1323	1423	1523	1623	1723	1823	1923	2023	2123	2223	2335
East Grinstead	arr	0018	0828	0928	1028	1128	1228	1328	1428	1528	1628	1728	1828	1928	2028	2128	2228	2340
Edenbridge Town	dep	....	....	....	....	....	....	....	....	....	....	....	....	....	....	....	....	....
Hever	dep	....	....	....	....	....	....	....	....	....	....	....	....	....	....	....	....	....
Cowden	dep	....	....	....	....	....	....	....	....	....	....	....	....	....	....	....	....	....
Ashurst	dep	....	....	....	....	....	....	....	....	....	....	....	....	....	....	....	....	....
Eridge	dep	....	....	....	....	....	....	....	....	....	....	....	....	....	....	....	....	....
Crowborough	dep	....	....	....	....	....	....	....	....	....	....	....	....	....	....	....	....	....
Buxted	dep	....	....	....	....	....	....	....	....	....	....	....	....	....	....	....	....	....
Uckfield	arr	....	....	....	....	....	....	....	....	....	....	....	....	....	....	....	....	....

Notes b – Change at East Croydon
u – Stops to pick up only.

30 September 2001 to 1 June 2002

SouthCentral

TRAIN TIMES

Beware: most video recorders use the 24-hour clock!

To avoid possible confusion, most timetables are written using the 24-hour clock. Otherwise we could mistake 7 a.m. for 7 p.m.

Example Change these times to 24-hour clock times.

a 7 a.m. **b** 3 p.m. **c** 6.45 p.m.

a 7 a.m. is written as 0700.
b 3 p.m. is written as 1500.
c 6.45 p.m. is written as 1845.

Remember:
To change p.m. times to 24-hour clock times add 12 hours.
To change 24-hour clock times that are after 1200 to a 12-hour clock, subtract 12 hours.

Exercise 3.1

1 Change these times to 24-hour clock times.
  a 2.30 p.m.   **b** 9 p.m.   **c** 8.45 a.m.   **d** 6 a.m.
  e Midday   **f** 10.55 p.m.   **g** 7.30 a.m.   **h** 7.30 p.m.
  i 1 a.m.   **j** Midnight

2 Change these times into those on the 24-hour clock.
  a A quarter past seven in the morning
  b Eight o'clock at night
  c Ten past nine in the morning
  d A quarter to nine in the morning
  e A quarter to three in the afternoon
  f Twenty to eight in the evening

3 These times are written for the 24-hour clock. Rewrite them using a.m. and p.m.
  a 0720   **b** 0900   **c** 1430   **d** 1825
  e 2340   **f** 0115   **g** 0005   **h** 1135
  i 1750   **j** 2359   **k** 0410   **l** 0545

4 A bus journey to work takes a woman three quarters of an hour. If she catches the bus at the following times, when does she arrive?
  a 0720   **b** 0755   **c** 0820   **d** 0845

5 The bus journey home from work takes the same woman 55 minutes. If she catches the bus at the following times, when does she arrive?
  a 1725   **b** 1750   **c** 1805   **d** 1820

6 A boy cycles to school each day. His journey takes 70 minutes. When will he arrive if he leaves home at:
  a 0715   **b** 0825   **c** 0840   **d** 0855

7 The train journey into the city from a village takes 1 hour 40 minutes. Copy and complete the train timetable below.

depart	arrive
0615	0755
	0810
0925	
	1200
1318	
	1628
1854	
	2103

8 The bus journey from the same village into the city takes 2 hours 5 minutes. Copy and complete the bus timetable below.

depart	arrive
0600	
	0850
0855	
	1114
1348	
	1622
2125	
	0010

9 A coach runs from Cambridge to airports at Stansted, Gatwick and Heathrow. The intended time taken for the journey remains constant. Copy and complete the timetables below for outward and return journeys.

Cambridge	0400	0835	1250	1945	2110
Stansted	0515				
Gatwick	0650				
Heathrow	0835				

Heathrow	0625	0940	1435	1810	2215
Gatwick	0812				
Stansted	1003				
Cambridge	1100				

10 British Airways aircraft fly twice a day from London to Johannesburg in South Africa. The flight time is 11 hours 20 minutes. Copy and complete the timetable below.

	London	Jo'burg	London	Jo'burg
Sunday	0615		1420	
Monday		1843		0525
Tuesday	0720		1513	
Wednesday		1912		0730
Thursday	0610		1627	
Friday		1725		0815
Saturday	0955		1850	

Speed, distance and time

You will already be familiar with the following formula:

$$\text{distance} = \text{speed} \times \text{time}$$

Rearranging the formula gives:

$$\text{time} = \frac{\text{distance}}{\text{speed}} \quad \text{and} \quad \text{speed} = \frac{\text{distance}}{\text{time}}$$

Where the speed is not constant:

$$\text{average (mean) speed} = \frac{\text{total distance}}{\text{total time}}$$

Remember:

Exercise 3.2

Remember:
Always state units in your answers.

1 Find the average (mean) speed of an object that travels:
   a 30 m in 5 seconds
   c 78 m in 2 hours
   e 400 km in 2 hours and 30 min
   b 48 m in 12 seconds
   d 50 km in 2.5 hours
   f 110 km in 2 h 12 min

2 How far will an object travel during:
   a 10 s at 40 m/s
   c 3 hours at 70 km/h
   e 10 mins at 60 km/h
   b 7 s at 26 m/s
   d 4 h 15 min at 60 km/h
   f 1 h 6 min at 20 m/s

3 How long will an object take to travel:
   a 50 m at 10 m/s
   c 2 km at 30 km/h
   e 200 cm at 0.4 m/s
   b 1 km at 20 m/s
   d 5 km at 70 m/s
   f 1 km at 15 km/h

The graph of distance against time for an object travelling at a constant speed is a straight line, as shown below.

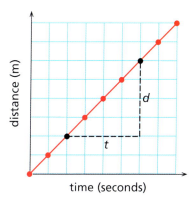

$$\text{gradient} = \frac{d}{t}$$

The units of the gradient are m/s, hence the gradient of a distance–time graph represents the speed at which the object is travelling.

Examples   The graph below represents an object travelling at constant speed.

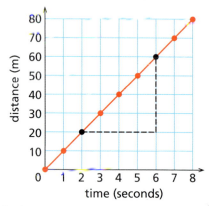

a From the graph, calculate how long it took the object to cover a distance of 30 m.
b Calculate the gradient of the graph.
c Calculate the speed at which the object was travelling.

a The time taken to travel 30 m is 3 seconds.
b Taking two points on the line as shown, the gradient $= \frac{40}{4} = 10$.
c On a distance–time graph, the speed is given by the gradient of the graph. Therefore the speed is 10 m/s.

Exercise 3.3

You will need:
• graph paper

1 Draw a distance–time graph for the first 10 seconds of a journey of an object travelling at a constant speed of 6 m/s.
2 Draw a distance–time graph for the first 10 seconds of a journey of an object travelling at 5 m/s. Use your graph to estimate:
  a the time taken to travel 25 m
  b how far the object travels in 3.5 seconds.
3 Two objects A and B set off from the same point and travel in the same direction. B sets off first, while A sets off 2 seconds later as shown below.

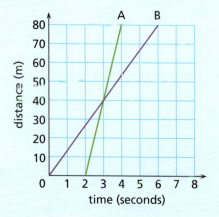

Using the graph, calculate:
a the speed of each of the objects A and B
b how far apart the objects would be 20 seconds after the start.

4 Three objects A, B and C move in the
same direction away from a point X.
Both A and C change their speed during
the journey, while B travels at a constant
speed throughout. A distance–time graph
of their journeys is shown below.
From the distance–time graph, estimate:
a the speed of object B
b the two speeds of object A
c the average speed of object C
d how far object C is from X,
3 seconds from the start
e how far apart objects A and C
are 4 seconds from the start.

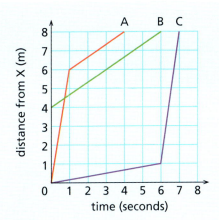

Travel graphs

The graphs of journeys of two or more objects can be shown on one grid with
the same axes. The shape of each graph gives a clear picture of the movement of
that object.

Example   Car X and car Y both reach point B 100 km from A at 11 a.m.

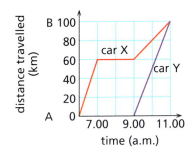

a Calculate the speed of car X between 6 a.m. and 7 a.m.
b Calculate the speed of car Y between 9 a.m. and 11 a.m.
c Explain what is happening to car X between 7 a.m. and 9 a.m.

a Speed $= \dfrac{\text{distance}}{\text{time}}$

$\quad = \dfrac{60}{1}\,\text{km/h}$

$\quad = 60\,\text{km/h}$

b Speed $= \dfrac{100}{2}\,\text{km/h}$

$\quad = 50\,\text{km/h}$

c No distance has been travelled, therefore car X is stationary.

Exercise 3.4

You will need:
• graph paper

1 Two friends, Paul and Helena, arrange to meet for lunch at noon. They live 50 km apart and the restaurant is 30 km from Paul's home. The travel graph below illustrates their journeys.

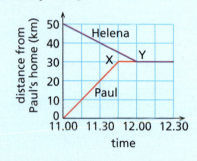

    a What is Paul's average speed between 11 a.m. and 12 noon?
    b What is Helena's speed between 11 a.m. and noon?
    c What does the line XY represent?

2 A car travels at a speed of 60 km/h for 1 hour. It stops for 30 minutes, then continues at a constant speed of 80 km/h for a further $1\frac{1}{2}$ hours. Draw a distance–time graph for this journey.

3 A girl cycles for $1\frac{1}{2}$ hours at 10 km/h. She stops for an hour, then travels a further 15 km in 1 hour. Draw a distance–time graph for the girl's journey.

4 Two friends, Theodore and Yin, leave their houses at 4 p.m. The houses are 4 km apart and the friends travel towards each other on the same road. Theodore walks at 7 km/h and Yin at 5 km/h.

    a On one grid and set of axes, draw a distance–time graph for their journeys.
    b From your graph, estimate the time at which the friends meet.
    c Estimate the distance from Theodore's house to the point where the friends meet.

Spending and saving

Specimen credit cards

Exercise 3.5

A	B	C	D
What are: • current accounts? • deposit accounts? • bank statements? • debit cards?	What is the difference between these? • hire purchase • overdraft • bank loan • student loan What is APR?	Is it better to rent a place to live or to buy a place to live by taking out a mortgage on a property?	Credit cards seem easy to get. What are the advantages and disadvantages of having a credit card?

Working in small groups, choose one of the discussion boxes A–D above. Discuss the questions, then give a presentation explaining the facts and opinions from your discussion.

Your presentation could be given in one of the following ways:

• as a children's educational programme for children of primary school age
• as a party political broadcast
• as an advertisement for 18 year olds
• as a quiz show.

Getting paid

Net pay is what is left after deductions such as tax, insurance and pension contributions are taken from **gross earnings**.

Trustco's Supermarkets	Pay advice			Week ending 31-8-2002
Payee no.	Tax year	Employee's name	Tax code	N.I. number
154	2002/2003	DAN JACKSON	347L	NW90480B

		Code	Value	Tax	£42.05
Basic pay	£245.00				
Overtime	£27.57			N.I.	£19.68
Total deductions	£73.98	123	£245.00		
NET PAY	£198.59	124	£27.57	Pens. contr.	£12.25

net pay = gross pay − deductions

A **bonus** is an extra payment sometimes added to an employee's basic pay.

In many jobs there is a fixed number of hours that the employee is expected to work. Any work done over this **basic week** is paid at a higher rate, referred to as **overtime**. Typical overtime may be 1.5 times basic pay, called **time and a half**, or it may be twice basic pay, called **double time**.

Exercise 3.6

1 Copy the table below and find the net pay for the four employees.

		gross pay (£)	deductions (£)	net pay (£)
a	A. Ahmet	162.00	23.50	
b	B. Martinez	205.50	41.36	
c	C. Stein	188.25	33.43	
d	D. Wong	225.18	60.12	

2 Copy and complete the table below for the five employees.

		basic pay (£)	overtime (£)	bonus (£)	gross pay (£)
a	P. Small	144	62	23	
b	B. Smith	152		31	208
c	A. Chang		38	12	173
d	U. Zafer	115	43		213
e	M. Said	128	36	18	

3 Copy and complete the table below for the four employees.

		gross pay (£)	tax (£)	pension (£)	net pay (£)
a	A. Hafar	203	54	18	
b	K. Zyeb		65	23	218
c	H. Such	345		41	232
d	K. Donald	185	23		147

4 Find the gross pay in each of the cases below. Copy and complete the table.

	no. of hours worked	basic rate per hour (£)	gross pay (£)
a	40	4.15	
b	44	4.88	
c	38	5.02	
d	35	8.30	
e	48	7.25	

5 Copy and complete the table below, which shows basic pay and overtime at time and a half.

	basic hours worked	rate per hour (£)	basic pay (£)	overtime hours worked	overtime pay (£)	total gross pay (£)
a	40	4.60		8		
b	35		203.00	4		
c	38	4.15		6		
d		6.10	256.20	5		
e	44	5.25		4		
f		4.87	180.19	3		
g	36	6.68		6		
h	45	7.10	319.50	7		

Simple interest

Interest is money added by a bank or building society to sums deposited by customers, or it is money charged by a bank or building society to customers for borrowing. The money deposited or borrowed is called the **principal**. The **percentage interest** is the given rate and the money is usually left or borrowed for a fixed period of time.

A formula can be used for calculating **simple interest**:

$$SI = \frac{PTR}{100}$$

where **SI** is the interest paid
P is the principal (the amount borrowed or lent)
T is the time in years
R is the rate percentage.

Examples Find the simple interest earned on £250 deposited for six years at 8% per annum (p.a.).

$$SI = \frac{PTR}{100}$$

$$= \frac{250 \times 6 \times 8}{100}$$

$$= 120$$

So the interest paid is £120.

How long will it take for a sum of £250 invested at 8% per annum to earn interest of £80?

$$SI = \frac{PTR}{100}$$

$$80 = \frac{250 \times T \times 8}{100}$$

$$8000 = 250 \times T \times 8$$
$$8000 = 2000 \times T, \text{ i.e. } T = 4$$

Therefore the time is four years.

....

What rate per year must be paid for a principal of £750 to earn interest of £180 in four years?

$$SI = \frac{PTR}{100}$$

$$180 = \frac{750 \times 4 \times R}{100}$$

$$180 = 30 \times R, \text{ i.e. } R = 6$$

Therefore the rate per year is 6%.

Exercise 3.7

All rates of interest are annual rates.

1 Find the simple interest paid in the following cases.
  a Principal £300   rate 6%   time 4 years
  b Principal £750   rate 8%   time 7 years
  c Principal £425   rate 6%   time 4 years
  d Principal £2800   rate 4.5%   time 2 years
  e Principal £880   rate 6%   time 7 years

2 How long will it take for the interest to be earned in the following cases?
  a $P = £500$ $R = 6\%$ $SI = £150$
  b $P = £5800$ $R = 4\%$ $SI = £96$
  c $P = £4000$ $R = 7.5\%$ $SI = £1500$
  d $P = £2800$ $R = 8.5\%$ $SI = £1904$
  e $P = £900$ $R = 4.5\%$ $SI = £243$
  f $P = £400$ $R = 9\%$ $SI = £252$

3 Calculate the rate of interest per year which will earn the given amount of interest in each of the following cases.
  a Principal £400   time 4 years   interest £112
  b Principal £800   time 7 years   interest £224
  c Principal £2000   time 3 years   interest £210
  d Principal £1500   time 6 years   interest £675
  e Principal £850   time 5 years   interest £340
  f Principal £1250   time 2 years   interest £275

4 Calculate the principal that will earn the interest stated, in the given number of years and at the given rate.
  a $SI = £80$ time 4 years   rate $= 5\%$
  b $SI = £36$ time 3 years   rate $= 6\%$
  c $SI = £340$ time 5 years   rate $= 8\%$
  d $SI = £540$ time 6 years   rate $= 7.5\%$
  e $SI = £540$ time 3 years   rate $= 4.5\%$
  f $SI = £348$ time 4 years   rate $= 7.25\%$

5 What rate of interest is paid on a deposit of £2000 that earns £400 interest in five years?
6 How long will it take a principal of £350 to earn £56 interest at 8% per year?
7 A principal of £480 earns £108 interest in five years. What rate of interest was being paid?
8 A principal of £750 becomes a total of £1320 in eight years. What rate of interest was being paid?
9 £1500 is invested for six years at 3.5% per year. What is the interest earned?
10 £500 is invested for 11 years and becomes £830 in total. What rate of interest was being paid?

Compound interest

In the problems on simple interest, interest is paid only on the amount deposited, the principal. When interest is also paid on interest, it is said to be **compounded**. **Compound interest** sounds complicated but the following example should make it clear.

A builder is going to build six houses on a plot of land. He borrows £500 000 at 10% interest and will pay off the loan in full after three years.

His total payment with compound interest is worked out as follows.

> **Remember:**
> If you **deposit** money, you earn interest. If you **borrow** money, you are charged interest.

At the end of year 1, his debt will be
  £500 000 + 10% of £500 000,
  i.e. £550 000
At the end of year 2, his debt will be
  £550 000 + 10% of £550 000,
  i.e. £550 000 + £55 000
  = £605 000
At the end of year 3, his debt will be
  £605 000 + 10% of £605 000,
  i.e. £605 000 + £60 500
  = £665 500

His interest is £665 500 − £500 000 = £165 500

Note. At simple interest, his interest would have been only £150 000 (£50 000 per year).

Exercise 3.8

1a A shipping company borrows £70 million at 5% compound interest to build a new cruise ship. If it repays the debt after three years, how much interest will the company pay?
  b What would have been the simple interest on the same £70 million loan for three years?
2 A woman takes a £100 000 mortgage on a property she is renovating. The interest rate is 15%. How much interest will she pay if she repays the mortgage after three years?
3 A man owes £6000 on a credit card. His annual percentage rate (APR) of interest is $33\frac{1}{3}\%$. If he lets the debt grow, how much will he owe after three years?

SUMMARY

By the time you have completed this chapter you should know:

■ how to understand and use the 12-hour and 24-hour clocks, for example

   3.30 p.m. is 1530 using the 24-hour clock

■ how to use timetables involving the 24-hour clock
■ how to use speed, distance and time in calculations
■ how to interpret and display information on travel graphs
■ the major financial terms such as loan, current account, deposit account, overdraft, hire purchase, credit card, rent, mortgage, wage, salary, gross pay, net pay, tax
■ how to carry out calculations involving **simple interest** (SI) and **compound interest**.

$$SI = \frac{PTR}{100} \quad \text{where } P \text{ is the principal, } T \text{ is the time and } R \text{ is the rate}$$

Exercise 3A

1 Change the times below into those on the 24-hour clock.
   a 4.35 a.m.
   b 6.30 p.m.
   c Quarter to 8 in the morning
   d Half past seven in the evening

2 The times below are written for the 24-hour clock. Rewrite them using a.m. and p.m.
   a 0845   b 1835   c 2112   d 0015

3 A journey to school takes a girl 25 minutes. What time does she arrive if she leaves home at the following times?
   a 0745   b 0815   c 0838

4 A bus service visits the towns on this timetable. Copy the timetable and fill in the missing times, given the information below the table.

Alphaville	0750		
Betatown		1138	
Gammatown			1648
Deltaville			

The journey from:
Alphaville to Betatown takes 37 minutes
Betatown to Gammatown takes 18 minutes
Gammatown to Deltaville takes 42 minutes

5 Find the time taken for each of the following journeys, given the distance and the average speed. Give your answers in hours and minutes.
   a 250 km at 50 km/h
   b 375 km at 100 km/h
   c 80 km at 60 km/h
   d 200 km at 120 km/h

6 A boy worked 3 hours a day each weekday for £4.15 per hour. What is his weekly gross payment?

7 A woman works at home making curtains. Write an algebraic formula for her gross pay if she receives £2.10 for each long curtain and £1.85 for each short curtain.

8 Copy and complete the simple interest table below.

	principal (£)	rate (%)	time (years)	interest (£)
a	200	9	3	
b	350	7		98
c	520		6	169
d		3.75	6	189

Exercise 3B

1 Change the times below to those on the 24-hour clock
   a 5.20 a.m.   **b** 8.15 p.m.
   c Ten to nine in the morning   **d** Half past eleven at night
2 The times below are written for the 24-hour clock. Rewrite them using a.m. and p.m.
   a 0715   **b** 1643   **c** 1930   **d** 0035
3 A journey to school takes a boy 22 minutes. When does he arrive if he leaves at the following times?
   a 0748   **b** 0817   **c** 0838
4 A train stops at the following stations. Copy the timetable and fill in the missing times, given the information below the table.

Apple	1014		
Peach		1720	
Pear			2315
Plum			

The journey from:
   Apple to Peach is 1 hour 38 minutes
   Peach to Pear is 2 hours 4 minutes
   Pear to Plum is 1 hour 53 minutes
5 Find the time taken for each of the following journeys, given the distance and the average speed. Give your answers in hours and minutes.
   a 350 km at 70 km/h
   b 425 km at 100 km/h
   c 160 km at 60 km/h
   d 450 km at 120 km/h
6 A girl works in a shop on Saturdays for $8\frac{1}{2}$ hours. She is paid £4.50 per hour. What is her gross pay for four weeks' work?
7 A potter makes cups and saucers in a factory. He is paid £14.40 per dozen cups and £12.00 per dozen saucers. What is his gross pay if he makes 9 dozen cups and 11 dozen saucers in one day?
8 Copy and complete the simple interest table below.

	principal (£)	rate (%)	time (years)	interest (£)
a	300	6	4	
b	250		3	60
c	480	5		96
d	650		8	390
e		3.75	4	187.50

Exercise 3C

A family are taking a touring holiday in Scotland and decide to hire a large people carrier. They will need the vehicle for 12 days and expect to cover about 3000 km.

They obtain three quotes for roughly similar vehicles.

a Which company should they choose based on the costs shown?

Fuel costs 80p per litre and the people carrier on average covers 15 km per litre. The family are also allowing a further £150 per day for food and accommodation.

b What is the *total* cost of the holiday based on the figures quoted?

Abbey Rental
£55 per day unrestricted travel

Barker Hire
£30 per day plus 7p per kilometre

Cost Less
£40 per day, 1500km free of charge 10p per kilometre afterwards

Exercise 3D

Using a spreadsheet, model the 12-day holiday to Scotland described in exercise 3C. Make sure your spreadsheet includes the following:

- a graph showing the costs of hiring the people carrier from the three companies for a range of journey lengths up to 5000 km
- a clearly identified 'Total Cost' cell based on a tour of 3000 km

An example of how you may wish to set up your spreadsheet is shown below.

You will need:
- computer with spreadsheet package installed

	A	B	C	D	E	F	G	H	I	J
	Distance	Car Hire (£)			Fuel Consumption	Fuel Cost				
1					(litres)	(£)				
2	(km)	Abbey Rental	Barker Hire	Cost Less				Total Accommodation + Food (£)		
3	200									
4	400									
5	600							Grand Total (£)		
6	800									
7	1000									
8	etc									
9										
10										
11										
12		Enter formulae in each of these columns								
13										

Exercise 3E

Using the internet as a resource, find answers to the following questions.

- What was the Wall Street Crash?
- How did it contribute to the Great Depression of the 1930s?
- What was the New Deal?

4 Expressions, inequalities and equations

'The yield of 2 sheaves of superior grain, 3 sheaves of medium grain and 4 sheaves of inferior grain is each less than 1 *tou*. But if one sheaf of medium grain is added to the superior grain, or if one sheaf of inferior grain is added to the medium, or if 1 sheaf of superior grain is added to the inferior, then in each case the yield is exactly 1 *tou*.

What is the yield of one sheaf of each grade of grain?'

The problem above comes from the most important book of ancient Chinese mathematics, called *Nine Chapters on the Mathematical Arts*. The book was written approximately 2000 years ago. As its name suggests, it consists of nine chapters. Each chapter presents a series of mathematical problems related to life in China at the time.

Chapter 8 deals with problems involving the solution of simultaneous equations.

> **Remember:**
> *The basic rule for solving equations is to do the same to both sides.*

In *Intermediate 1* you learned:

- what is meant by an equation
- how to solve linear equations
- how to solve linear equations involving brackets
- how to solve linear equations involving negatives
- how to construct a linear equation.

Expressions

An **expression** is used to represent a value in algebraic form. For example

The length of this line is given by the expression $x + 3$.

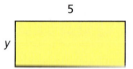

The perimeter of the rectangle is

$$y + 5 + y + 5 = 2y + 10$$

The area of the rectangle is $5y$.

Exercise 4.1

1 Write an expression for the perimeter of each of the following shapes.

> The diagrams in this chapter are not drawn to scale.

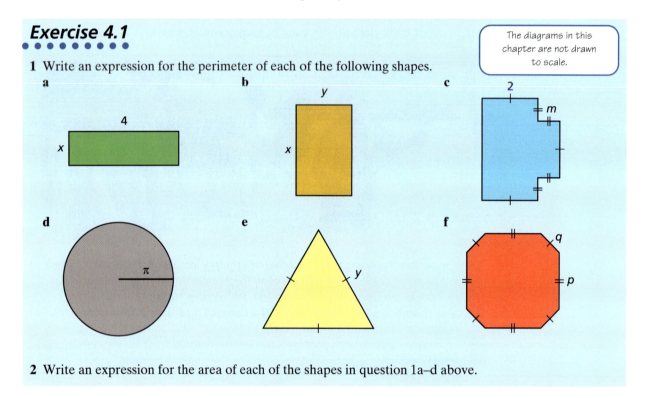

2 Write an expression for the area of each of the shapes in question 1a–d above.

Expanding two linear expressions

An expression for the area of the shape below is more complicated.

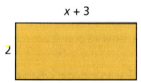

The rectangle can be split in two as follows:

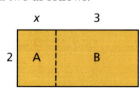

Area of A can be expressed as $2x$.
Area of B can be expressed as 6.
Total area of the rectangle is $2x + 6$.

The area of the rectangle can also be expressed using brackets as $2(x + 3)$.
Therefore

$$2(x + 3) = 2x + 6$$

Remember, that to **expand** brackets, multiply the terms in the brackets by the term outside, i.e. with $2(x + 3)$, the 2 multiplies both the x and the 3.

> **Remember:**
> *When you expand an expression you are getting rid of the brackets.*

$$2(x + 3) = 2 \times x + 2 \times 3$$
$$= 2x + 6$$

Example Write an expression for the area of the following rectangle:

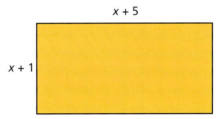

The rectangle can be split into four as follows:

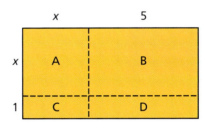

Area of A is $x \times x = x^2$.
Area of B is $5 \times x = 5x$.
Area of C is $1 \times x = x$.
Area of D is $1 \times 5 = 5$.

Total area is given by the expression

$$x^2 + 5x + x + 5$$
$$= x^2 + 6x + 5$$

Using brackets, the area of the rectangle can also be expressed as follows:

$$(x + 1)(x + 5)$$

Therefore $(x + 1)(x + 5) = x^2 + 6x + 5$.
   As before, to expand the product of brackets, multiply all the terms in one set of brackets by the terms in the other set of brackets.

> Since $(x + 1)(x + 5) = x^2 + 6x + 5$ is true for all values of x, then that equation is known as an **identity**.

$$(x + 1)(x + 5) = x^2 + 5x + x + 5$$
$$= x^2 + 6x + 5$$

Exercise 4.2

For each of the following shapes:

a write, using brackets, an expression for the area

b expand your expression from part **a** by multiplying out the brackets.

1
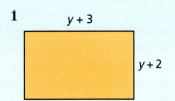
$y + 3$
$y + 2$

2
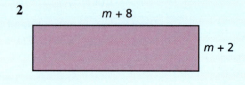
$m + 8$
$m + 2$

3
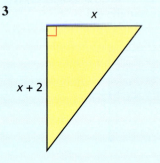
x
$x + 2$

4
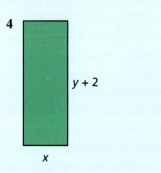
$y + 2$
x

5
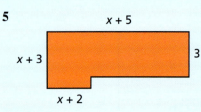
$x + 5$
$x + 3$
3
$x + 2$

6
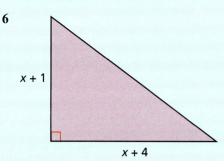
$x + 1$
$x + 4$

7
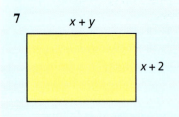
$x + y$
$x + 2$

8
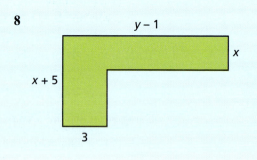
$y - 1$
x
$x + 5$
3

In questions **9** and **10** show that the equations are identities.

9 a $(3a + 2)(3a - 2) = 9a^2 - 4$
  b $(3a - 2)^2 = 9a^2 - 12a + 4$
10 a $(2a + 3b)(2a - 3b) = 4a^2 - 9b^2$
   b $(2a + 3b)^2 = 4a^2 + 12ab + 9b^2$

Exercise 4.3

Expand each of the following and simplify your answer.

1 a $(x - 2)(x + 3)$ **b** $(x + 8)(x - 3)$
2 a $(x + 1)(x - 3)$ **b** $(x - 7)(x + 9)$
3 a $(x - 3)(x - 3)$ **b** $(x - 7)(x - 5)$
4 a $(a + b)(a - b)$ **b** $(p - q)(p + q)$
5 a $(3y + 4)(2y + 5)$ **b** $(6y + 3)(3y + 1)$
6 a $(6p + 2)(3p - 1)$ **b** $(-7p - 3)(4p - 8)$
7 a $(2x + 6)^2$ **b** $(2x + 3)(2x - 3)$
8 a $(-3 - 2y)(4y - 6)$ **b** $(7 - 5y)^2$

Inequalities

With equations we are used to the unknown variable having set values so that one side of the equation is *equal* to the other, for example $4x = 8$.

With an **inequality**, however, the variable can take a range of values. The symbols used with inequalities are as follows:

> $>$ means 'is greater than'
> $\geqslant$ means 'is greater than or equal to'
> $<$ means 'is less than'
> $\leqslant$ means 'is less than or equal to'

> $x \geqslant 3$ states that x is greater than or equal to 3,
> i.e. x can be 3, 4, 4.2, 5, 5.6, etc.
> $3 \leqslant x$ states that 3 is less than or equal to x,
> i.e. x is still 3, 4, 4.2, 5, 5.6, etc.

Therefore:

$5 > x$ can be rewritten as $x < 5$,
i.e. x can be 4, 3.2, 3, 2.8, 2, 1, etc.
$-7 \leqslant x$ can be rewritten as $x \geqslant -7$,
i.e. x can be $-7, -6, -5$, etc.

These inequalities can also be represented on a number line:

> If $x \geqslant 3$ then x can be 3 or 6 or 824 or 3×10^{23} or..., i.e. 3 or any number more than 3.

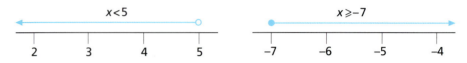

Note that ○ implies that the number is not included in the solution, while ● implies that the number is included in the solution.

Examples The maximum number of players (n) from one football team allowed on the pitch at any one time is 11. Show this information:
a as an inequality **b** on a number line.

a $n \leqslant 11$
b

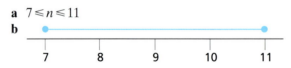

The maximum number of players (n) from one football team allowed on the pitch at any one time is 11. The minimum number allowed is seven players. Show this information:
a as an inequality **b** on a number line.

a $7 \leqslant n \leqslant 11$
b

Exercise 4.4

1 Copy each of the following statements and insert one of the symbols =, >, < into the space, to make the statement correct.
a $7 \times 2 \ldots 8 + 7$ **b** $6^2 \ldots 9 \times 4$
c $5 \times 10 \ldots 7^2$ **d** $80\,\text{cm} \ldots 1\,\text{m}$
e $1000\,\text{litres} \ldots 1\,\text{m}^3$ **f** $48 \div 6 \ldots 54 \div 9$

> **Remember:**
> ● means the number is included.
> ○ means the number isn't included.

2 Represent each of the following inequalities on a number line.
a $x < 2$ **b** $x \geqslant 3$
c $x \leqslant -4$ **d** $x \geqslant -2$
e $2 < x < 5$ **f** $-3 < x < 0$
g $-2 \leqslant x < 2$ **h** $2 \geqslant x \geqslant -1$

3 Write down the inequality that corresponds to each of the following number lines.

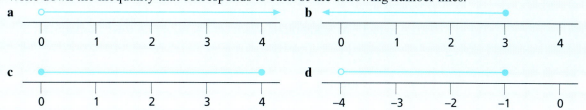

4 Write the following sentences using inequality signs.
  a The maximum capacity of an athletics stadium is 20 000 people.
  b In a class the tallest pupil is 180 cm and the shortest is 135 cm.
  c Five times a number plus 3 is less than 20.
  d The maximum temperature in May was 25 °C.
  e A farmer has more than 350 apples but fewer than 400 apples on each tree in his orchard.

> **Remember:**
> *Choose a letter,*
> *e.g. m for maximum,*
> *t for tallest, etc.*

Solving inequalities

Solving inequalities can, in many cases, be handled like solving an equation, i.e. what you do to one side of the inequality you also do to the other. However, there are some exceptions to this rule as shown below.

$6 < 8$ this inequality is true and can be manipulated in the following ways:

adding 2 to each side	$8 < 10$	this inequality is still true
subtracting 2 from each side	$4 < 6$	this inequality is still true
multiplying both sides by 2	$12 < 16$	this inequality is still true
dividing both sides by 2	$3 < 4$	this inequality is still true
multiplying both sides by -2	$-12 < -16$	this inequality is not true
dividing both sides by -2	$-3 < -4$	this inequality is not true

> **Remember:**
> *Change sign, change symbol:*
> $-5 < -2$
> $5 > 2$

As can be seen, when both sides of an inequality are either multiplied or divided by a negative number, the inequality is no longer true. For it to be true, the inequality sign needs to be changed around, i.e.

$$-12 > -16 \quad \text{and} \quad -3 > -4$$

Examples   Solve the following inequalities.
  a $15 + 3x < 6$ **b** $17 \leqslant 7x + 3$ **c** $9 - 4x \geqslant 17$

  a $15 + 3x < 6$
        $3x < -9$ (15 subtracted from each side)
         $x < -3$ (each side divided by 3)
  b $17 \leqslant 7x + 3$
       $14 \leqslant 7x$ (3 subtracted from each side)
       $2 \leqslant x$ or $x \geqslant 2$ (each side divided by 7)
  c $9 - 4x \geqslant 17$
      $-4x \geqslant 8$ (9 subtracted from each side)
        $x \leqslant -2$ (each side divided by -4)

Note. The inequality sign has changed direction.

Find the range of values for which $7 < 3x + 1 \leqslant 13$ and illustrate your solutions on a number line.

This is in fact two inequalities which can therefore be solved separately.

$$7 < 3x + 1 \quad \text{and} \quad 3x + 1 \leqslant 13$$
$$6 < 3x \qquad\qquad 3x \leqslant 12$$
$$2 < x, \text{ that is } x > 2 \qquad x \leqslant 4$$

Exercise 4.5

Solve each of the following inequalities and illustrate your solution on a number line.

1 a $x + 3 < 7$ **b** $5 + x > 6$ **c** $4 + 2x \leqslant 10$
   d $8 \leqslant x + 1$ **e** $5 > 3 + x$ **f** $7 < 3 + 2x$

2 a $x - 3 < 4$ **b** $x - 6 \geqslant -8$ **c** $8 + 3x > -1$
   d $5 \geqslant -x - 7$ **e** $12 > -x - 12$ **f** $4 \leqslant 2x + 10$

3 a $4 < 2x \leqslant 8$ **b** $3 \leqslant 3x < 15$ **c** $7 \leqslant 2x < 10$
   d $10 \leqslant 5x < 21$ **e** $12 < 8x - 4 < 20$ **f** $15 \leqslant 3(x - 2) < 9$

Graphing inequalities

The solution to an inequality can also be illustrated on a graph.

Examples   On a pair of axes, shade the region which satisfies the inequality $x \geqslant 3$.

To do this, draw the line $x = 3$. The region to the right of $x = 3$ represents the inequality $x \geqslant 3$ and therefore is shaded as shown below.

> Note: the coordinates of any point in the shaded region have an x-value > 3.

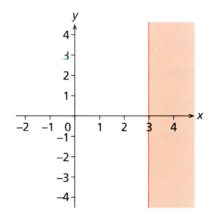

On a pair of axes, shade the region which satisfies the inequality $y < 5$.

<table>
<tr><td>

A dashed line means that the points on the line are *not* included in the range of results. A solid line means that the points on the line *are* included in the range of results.

</td></tr>
</table>

Draw the line $y = 5$ first (in this case draw it as a dashed line). The region below the line $y = 5$ represents the inequality $y < 5$ and is therefore shaded as shown.

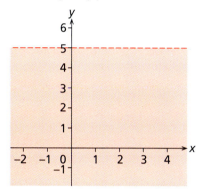

On a pair of axes, shade the region which satisfies the inequality $y \leqslant x + 2$.

Draw the line $y = x + 2$ first (since it is included in the solution, draw it as a solid line). To know which region satisfies the inequality, and therefore to know which side of the line to shade, follow these steps:

- choose a point at random which does not lie on the line, for example $(3, 2)$
- substitute those values of x and y into the inequality, i.e. $2 \leqslant 3 + 2$
- if the inequality is true, then the region the point lies in satisfies the inequality and can therefore be shaded.

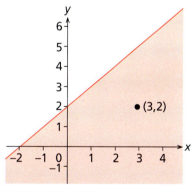

Note. In some questions, the region which satisfies the inequality is left *unshaded* while in others it is *shaded*. It is therefore important to read the question carefully to see which is wanted.

Exercise 4.6

You will need:
- squared paper

1 By drawing appropriate axes, shade the region which satisfies each of the following inequalities.

   a $x < 3$ **b** $y \leqslant 4$

   c $y > 2x + 1$ **d** $y \leqslant x - 3$

2 By drawing appropriate axes, leave *unshaded* the region which satisfies each of the following inequalities.

   a $y \geqslant -x$ **b** $y \leqslant 2 - x$

   c $2x - y \geqslant 3$ **d** $2y - x < 4$

> Shaded or unshaded? Read the question carefully!

Example   On the *same* pair of axes, leave *unshaded* the regions which satisfy the following inequalities.

$$x \leqslant 2 \quad y > -1 \quad y \leqslant 3 \quad y \leqslant x + 2$$

As a result, find the region which satisfies all four inequalities simultaneously.

If the four inequalities are graphed on separate axes, the solutions are as shown below.

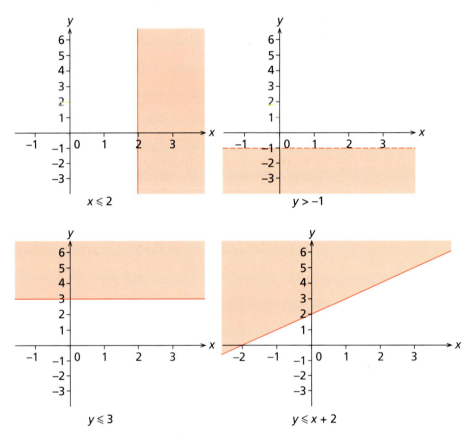

Combining all four on one pair of axes gives the following graph:

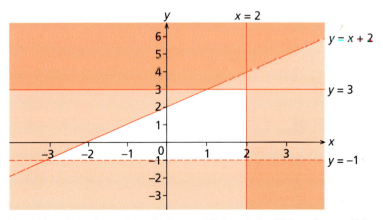

The unshaded region gives the solutions which satisfy all four inequalities.

Exercise 4.7

For each question, plot, on the same pair of axes, all the inequalities given. Leave *unshaded* the region which satisfies all of them simultaneously.

You will need:
• squared paper

1 $y \leqslant x$ $\quad$ $y > 1$ $\quad$ $x \leqslant 5$
2 $x + y \leqslant 6$ $\quad$ $y < x$ $\quad$ $y \geqslant 1$
3 $y \geqslant 3x$ $\quad$ $y \leqslant 5$ $\quad$ $x + y > 4$
4 $2y \geqslant x + 4$ $\quad$ $y \leqslant 2x + 2$ $\quad$ $y < 4$ $\quad$ $x \leqslant 3$

Solving simultaneous equations by elimination

So far we have looked at solving inequalities simultaneously (**simultaneous equations**). Equations can be solved simultaneously too. When you need to find the values of two unknowns, two equations need to be solved. The process of solving two equations and finding a solution which satisfies both of them is known as solving equations simultaneously. The method we are going to use here is called solving by **elimination**.

Example Solve the following equations simultaneously by finding the values of x and y which satisfy both equations.

$$3x + y = 9 \qquad (1)$$
$$5x - y = 7 \qquad (2)$$

By adding equations $(1) + (2)$, we can eliminate the variable y, giving:

$$8x = 16$$
$$x = 2$$

To find the value of y, substitute $x = 2$ into either equation (1) or equation (2). Substituting $x = 2$ into equation (1) gives:

$$3x + y = 9$$
$$6 + y = 9$$
$$y = 3$$

To check that the solution is correct, substitute the values of x and y into equation (2). If it is correct, then the left-hand side of the equation will equal the right-hand side.

$$5x - y = 7$$
$$10 - 3 = 7$$
$$7 = 7 \checkmark$$

Exercise 4.8

Solve the following simultaneous equations by elimination.

1 a $x + y = 6$
$\quad\quad x - y = 2$
b $5x + y = 29$
$\quad\, 5x - y = 11$
2 a $3x + 2y = 13$
$\quad\quad 4x = 2y + 8$
b $3x = 5y + 14$
$\quad\, 6x + 5y = 58$

3 a $2x + y = 14$
$\quad\quad x + y = 9$
b $x + 6y = -2$
$\quad\, 3x + 6y = 18$
4 a $x - y = 1$
$\quad\quad 2x - y = 6$
b $4x - y = -9$
$\quad\, 7x - y = -18$

> **Note.** In the expression 2x, the **coefficient** of x is the 2.

If neither x nor y can be eliminated by simply adding or subtracting the two equations then we need to multiply one or both of the equations. We multiply the equations by a number to make the **coefficients** of x (or y) equal.

Example The rectangle below has dimensions as shown. Calculate the values of x and y.

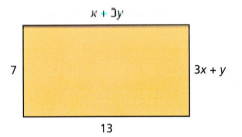

Two equations need to be formed:

$$x + 3y = 13 \qquad (1)$$
$$3x + y = 7 \qquad (2)$$

Here, simply adding or subtracting equation (1) and equation (2) will not eliminate either x or y. However, if we multiply equation (1) by 3 we get the following:

$$3x + 9y = 39 \qquad (3)$$
$$3x + y = 7 \qquad (2)$$

Subtracting equation (2) from (3) gives:

$$8y = 32$$

Therefore

$$y = 4$$

Substituting $y = 4$ into equation (1) gives:

$$x + 12 = 13$$

Therefore

$$x = 1$$

Substituting $x = 1$ and $y = 4$ into equation (2) to check gives:

$$3 + 4 = 7$$
$$7 = 7 \checkmark$$

Exercise 4.9

In questions 1–3, solve the pairs of simultaneous equations.

1 a $2x + y = 7$
$3x + 2y = 12$
    b $x + 5y = 11$
$2x - 2y = 10$

2 a $x + y = 5$
$3x - 2y + 5 = 0$
    b $x + y = 5$
$2x - 2y = -2$

3 a $3y = 9 + 2x$
$3x + 2y = 6$
    b $8y = 3 - x$
$3x - 2y = 9$

4 The sum of two numbers is 17 and their difference is 3. Find the two numbers by forming two equations and solving them simultaneously.

5 The difference between two numbers is 7. If their sum is 25, find the two numbers by forming two equations and solving them simultaneously.

6 Find the values of x and y.

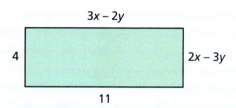

7 A man's age is three times his son's age. Ten years ago he was five times his son's age. By forming two equations and solving them simultaneously, find both of their ages.

8 A grandfather is ten times older than his grand-daughter. He is also 54 years older than her. How old are they?

SUMMARY

By the end of this chapter you should know:

■ what an **expression** is
■ how to **expand** the product of two linear expressions, for example

$$(x + 3)(y - 5) = xy - 5x + 3y - 15$$

■ what is meant by an **inequality** and the symbols involved

$$> \quad \geqslant \quad < \quad \leqslant$$

■ how to represent an inequality on a number line
■ how to solve inequalities such as

$$3x + 2 \leqslant 11$$

■ how to graph inequalities and be aware of when to use a dashed or solid line
■ how to solve simultaneous equations by **elimination**
■ that since

$$(x + 2)(x + 3) = x^2 + 5x + 6$$

is true for all values of x, then that equation is known as an **identity**

Exercise 4A

1 Write an expression for the area of this right-angled triangle.

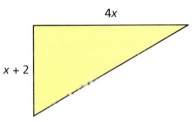

2 Expand the expressions below and simplify where possible.
  a $(4x + 3)(x - 2)$ **b** $(m + n)(m - n)$

3 Write the following sentences using inequalities.
  a No more than 52 people can be carried on a bus.
  b There are between 24 and 38 students in a class.

4 Illustrate each of the following inequalities on a number line.
  a $y > 5$ **b** $-3 \leqslant y < 0$

5 Solve the inequality $17 + 5x \leqslant 42$.

6 Solve the following pair of simultaneous equations by elimination.

$$6x + 5y = 62$$
$$4x - 5y = 8$$

7 Show that this equation is an identity.

$$(3x - 4)^2 = 9x^2 - 24x + 16$$

Exercise 4B

1 Write an expression for the area of this parallelogram.

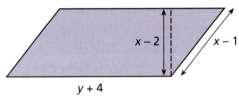

2 Expand the expressions below and simplify where possible.
  a $(2y - 3)(x + 3)$ **b** $(n - 4)(2n - 6)$

3 Write the information on the number lines below as inequalities.

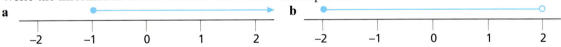

4 Illustrate each of the following inequalities on a number line.
  a $x \geqslant 3$ **b** $0 < x < 4$

5 Graph the inequality $y > x$. Shade the region which satisfies the inequality.

6 Solve the following pair of simultaneous equations.

$$4x + y = 14$$
$$6x - 3y = 3$$

7 Show that this equation is an identity.

$$(2x + 7)(3x - 1) = 6x^2 + 19x - 7$$

Exercise 4C

In the puzzle below, each of the fruits stands for a different number. The total of each row is written down the right-hand side.

By forming equations, and solving them, find the value of each fruit.

Exercise 4D

You will need:
• computer with graphing package installed

1 The number of fields a farmer plants with wheat is w and the number of fields he plants with corn is c. He plants no other crops. There are, however, certain restrictions on how many fields he can plant of each crop. These are as follows.

- There must be at least two fields of corn.
- There must be at least two fields of wheat.
- Not more than 10 fields in total are to be sown.

a Construct three inequalities from the information given above.
b Using a graphing package such as 'Omnigraph' or 'Autograph', graph the three inequalities on the same pair of axes. Leave *unshaded* the region which satisfies all three inequalities.
c From your graph, give two possible ways in which the farmer could plant his crops.

2 A taxi firm has at its disposal one morning a car and a minibus for hire. During the morning the firm makes x car trips and y minibus trips. The following facts are known.

- The firm makes at least five car trips.
- The firm makes between two and eight minibus trips.
- The total number of car and minibus trips does not exceed 12.

a Write an inequality for each of the statements above.
b Using a graphing package such as 'Omnigraph' or 'Autograph', graph the three inequalities on the same pair of axes. Leave *unshaded* the region which satisfies all three inequalities.
c From your graph, deduce a possible combination for the number of car and minibus trips that morning.

Exercise 4E

You will need:
- computer with internet access

Using the internet as a research tool, find out about the topics covered in the ancient Chinese book *Nine Chapters on the Mathematical Arts*. Try to solve some of the problems in chapter 8 of the book using modern mathematical methods.

5 Expressions, equations and formulae

Girolamo Cardano (1501–1576) was a famous Italian mathematician. In 1545 he published a book *Ars Magna* (Great Art) in which he showed calculations involving solutions of cubic equations (equations of the form $ax^3 + bx^2 + cx + d = 0$) and quartic equations ($ax^4 + bx^3 + cx^2 + dx + e = 0$).

His book (the title page is shown here) is one of the key historical texts on algebra. It was the first algebraic text written in Latin.

In 1570, because of his interest in astrology, Cardano was arrested for heresy; no other work of his was ever published.

Trial and improvement

Before Girolamo Cardano showed a way of solving equations such as

$$x^3 + 2x^2 + 4x + 5 = 0$$

an approximate solution could be reached by a process of **trial and improvement**. This involves trying different values for x to see how well they fit the equation, getting closer and closer to the solution.

For example, the equation $x^2 = 13$ has solution $\pm\sqrt{13}$. The value of $\sqrt{13}$ can be found easily using the $\boxed{\sqrt{}}$ key on a calculator. It can also be found by trial and improvement, as shown in the first example below.

Examples   Find, by trial and improvement, the solution to $x^2 = 13$, correct to 1 d.p.

If $x^2 = 13$, let us try different values for x, square them, and see how close to 13 we can get.

If $x = 3$, $x^2 = 9$ (too low)
If $x = 4$, $x^2 = 16$ (too high)

From the results above, it is clear that the solution to $x^2 = 13$ lies somewhere between $x = 3$ and $x = 4$. Therefore try some decimal values of x.

If $x = 3.5$ $x^2 = 12.25$ (too low)
If $x = 3.6$ $x^2 = 12.96$ (very close)
If $x = 3.7$ $x^2 = 13.69$ (too high)

To one decimal place, 3.6 is too low and 3.7 is too high. However, of the two solutions, 3.6 is the closer. Therefore $x = \pm 3.6$ to 1 d.p.

Find a positive value for x, correct to 1 d.p., if $x^2 + x = 104$.

$$\begin{aligned} &\text{If } x = 10 && x^2 + x = 100 + 10 = 110 && \text{(too high)} \\ &\text{If } x = 9 && x^2 + x = 81 + 9 = 90 && \text{(too low)} \end{aligned}$$

Therefore try decimal values of x between 9 and 10.

$$\begin{aligned} &\text{If } x = 9.5 && x^2 + x = 90.25 + 9.5 = 99.75 && \text{(too low)} \\ &\text{If } x = 9.6 && x^2 + x = 92.16 + 9.6 = 101.76 && \text{(too low)} \\ &\text{If } x = 9.7 && x^2 + x = 94.09 + 9.7 = 103.79 && \text{(close)} \\ &\text{If } x = 9.8 && x^2 + x = 96.04 + 9.8 = 105.84 && \text{(too high)} \end{aligned}$$

$104 - 103.79 = 0.21$
$105.84 - 104 = 1.84$
$0.21 < 1.84$

Therefore the positive solution to $x^2 + x = 104$ is $x = 9.7$ to 1 d.p.

Exercise 5.1

Use trial and improvement to find positive solutions for x in each of the following equations. Show your method clearly and give answers correct to 1 d.p.

1 $x^2 = 7$ 2 $x^2 = 19$ 3 $x^2 = 31$ 4 $x^2 + x = 13$
5 $x^2 - x = 25$ 6 $2x^2 = 27$ 7 $(2x)^2 = 150$ 8 $x^3 = 20$

Proof and counter-example

Girolamo Cardano would also have been familiar with the ideas of **verification**, **proof** and **counter-example**, which are used in the solution of equations.

Verification

The equation

$$3x + 1 = 5x - 7$$

can be solved to give $x = 4$. The verification (or check) is to substitute $x = 4$ in the equation.

$$\begin{aligned} 3x + 1 &= 5x - 7 \\ (3 \times 4) + 1 &= (5 \times 4) - 7 \\ 13 &= 13 \end{aligned}$$

So $x = 4$ in this particular case, but $3x + 1 \neq 5x - 7$ in *all* cases. A **proof** is needed to show that something is true in all cases.

Proof

Consider the equation

$$4(2x - 3y) = \frac{24x - 36y}{3}$$

If $x = 2$ and $y = 1$, then

$$4(4 - 3) = \frac{(24 \times 2) - (36 \times 1)}{3}$$

$$4 = 4 \qquad \text{This is a verification.}$$

But look more closely at the equation.

Left-hand side
$$4(2x - 3y)$$
$$= 8x - 12y$$

Right-hand side
$$\frac{24x - 36y}{3}$$
$$= \frac{24x}{3} - \frac{36y}{3}$$
$$= 8x - 12y$$

So the left-hand side = right-hand side, i.e.

$$8x - 12y = 8x - 12y$$

so the equation is proved true for *all* values of x and y.

Counter-example

If something is **proved** it must be true for all values and situations. So, if one single example is found of its not being true, a **counter-example**, then the proof collapses.

Consider the statement

'All positive whole numbers when doubled produce an even number.'

No counter-example has been found.

But consider the statement

'Squaring two numbers and then adding them together gives the same result as adding the two numbers and then squaring.'

Choose two numbers, say 2 and 3.

$$2^2 + 3^2 = 4 + 9 = 13$$
$$\text{but } (2 + 3)^2 = 5^2 = 25$$

This one counter-example is sufficient to show that the second statement can't be true.

Factorising

In chapter 4 you learned how to expand expressions like the ones below:

$$5x(2x - 3) = 10x^2 - 15x$$

$$(2x - 3)(x + 2) = 2x^2 + 2x - 3x - 6$$

$$= 2x^2 - x - 6$$

$$(x + 4)(2y - 3) = 2xy - 3x + 8y - 12$$

Exercise 5.2

(*Revision*)
Expand the following and simplify where possible.

1 $7(2x - 3)$
2 $4x(3x + 2)$
3 $5p(2q - 4)$
4 $x(2x - 3y)$
5 $(x - 4)(2y + 2)$
6 $(p - 3)(q - 2)$
7 $(x + 2)(x + 3)$
8 $(y - 4)(y - 5)$
9 $(m + 3)(m - 2)$
10 $(t + 5)(t - 5)$

In question 1 in exercise 5.2, $7(2x - 3)$ expanded to give $14x - 21$. The opposite of expanding is known as **factorising**.

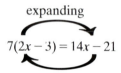

$$7(2x - 3) = 14x - 21$$

factorising

$14x - 21$ is factorised by looking at both terms (i.e. $14x$ and 21) and finding the **highest common factor** of both terms (i.e. the largest number that goes into both terms – in this case 7) and placing it outside the bracket.

Example Factorise the following expressions.
  a $10x + 15$
  b $8p - 6q + 10r$
  c $-2q - 6p + 12$
  d $2a^2 + 3ab - 5ac$
  e $6ax - 12ay - 18a^2$

  a $10x + 15$
  $= 5(2x + 3)$
  b $8p - 6q + 10r$
  $= 2(4p - 3q + 5r)$
  c $-2q - 6p + 12$
  $= 2(-q - 3p + 6)$
  d $2a^2 + 3ab - 5ac$
  $= a(2a + 3b - 5c)$
  e $6ax - 12ay - 18a^2$
  $= 6a(x - 2y - 3a)$

Note. In example **e**, $6ax - 12ay - 18a^2$ could be written as $6(ax - 2ay - 3a^2)$ but this is only *partly factorised* as a is still common to all the terms inside the bracket. When asked to factorise an expression, **complete factorisation** is needed.

Exercise 5.3

> Remember:
> *Factorise means complete factorisation is required.*

Factorise the following.

1 **a** $4x - 6$
   b $18 - 12p$
   c $6y - 3$
   d $4a + 6b$
   e $3p - 3q$
   f $8m + 12n + 16r$
2 **a** $3ab + 4ac - 5ad$
   b $8pq + 6pr - 4ps$
   c $a^2 - ab$
   d $4x^2 - 6xy$
   e $abc + abd + fab$
   f $3m^2 + 9m$

3 a $18 + 12y$
  d $4s - 16t + 20r$
4 a $m^2 + mn$
  d $ab + a^2b + ab^2$

b $14a - 21b$
e $5pq - 10qr + 15qs$
b $3p^2 - 6pq$
e $3p^3 - 4p^4$

c $11x + 11xy$
f $4xy + 8y^2$
c $pqr + qrs$
f $7b^3c + b^2c^2$

Factorisation by grouping

Expressions which don't have common factors in all their terms can sometimes be factorised by grouping. For example, if we are asked to factorise the expression $2xy + 4x + 3y + 6$, there does not appear to be a common factor. However, if the expression is split into two parts (groups) we can factorise each separately.

$$2xy + 4x + 3y + 6$$
$$\downarrow \qquad \downarrow$$
$$2x(y + 2) + 3(y + 2)$$

> Think of this as 2x lots of something and 3 lots of something, so there are 2x + 3 lots of something.

$(y + 2)$ is a common factor of both terms, so the expression can now be written as:

$$(2x + 3)(y + 2)$$

Example

Factorise the following expressions.
a $6x + 3 + 2xy + y$
b $ax + ay - bx - by$
c $2x^2 - 3x + 2xy - 3y$

> Equations such as
> $(a - b)(x + y)$
> $= ax + ay - bx - by$
> and
> $(a - b)(x + y)$
> $= a(x + y) - b(x + y)$
> which are true for all values of a, b, x, y are also known as **identities**.

a $6x + 3 + 2xy + y$
  $= 3(2x + 1) + y(2x + 1)$
  $= (3 + y)(2x + 1)$
Note that $(2x + 1)$ was a common factor of both terms.
b $ax + ay - bx - by$
  $= a(x + y) - b(x + y)$
  $= (a - b)(x + y)$
c $2x^2 - 3x + 2xy - 3y$
  $= x(2x - 3) + y(2x - 3)$
  $= (x + y)(2x - 3)$

Exercise 5.4

Factorise the following by grouping.

1 a $ax + bx + ay + by$
  c $3m + 3n + mx + nx$
2 a $pr - ps + qr - qs$
  c $ab - 4cb + ac - 4c^2$
3 a $xy + 4y + x^2 + 4x$
  c $ab + 3a - 7b - 21$
4 a $mn - 2m - 3n + 6$
  c $pr - 4p - 4qr + 16q$

b $ax + bx - ay - by$
d $4m + mx + 4n + nx$
b $pq - 4p + 3q - 12$
d $rs + rt + 2ts + 2t^2$
b $x^2 - xy - 2x + 2y$
d $ab - b - a + 1$
b $mn - 2mr - 3rn - 6r^2$
d $ab - a - bc + c$

Difference of two squares

On expanding $(x + y)(x - y)$ we get

$$(x + y)(x - y) = x^2 - xy + xy - y^2$$
$$= x^2 - y^2$$

The reverse is that $x^2 - y^2$ factorises to $(x + y)(x - y)$. Because x^2 and y^2 are both square, $x^2 - y^2$ is known as the **difference of two squares**.

Example   Factorise:

a $p^2 - q^2$ **b** $4a^2 - 9b^2$ **c** $(mn)^2 - 25k^2$ **d** $4x^2 - (9y)^2$

> **Remember:**
> After a factorisation, you can check your answer by expanding it.

a $p^2 - q^2$
$= (p + q)(p - q)$

b $4a^2 - 9b^2$
$= (2a)^2 - (3b)^2$
$= (2a + 3b)(2a - 3b)$

c $(mn)^2 - 25k^2$
$= (mn)^2 - (5k)^2$
$= (mn + 5k)(mn - 5k)$

d $4x^2 - (9y)^2$
$= (2x)^2 - (9y)^2$
$= (2x + 9y)(2x - 9y)$

Exercise 5.5

Factorise the following.

1 a $a^2 - b^2$
  d $m^2 - 49$
2 a $144 - y^2$
  d $1 - t^2$
3 a $9x^2 - 4y^2$
  d $x^2 - 100y^2$
4 a $m^2n^2 - 9y^2$
  d $4m^4 - 36y^4$

b $m^2 - n^2$
e $81 - x^2$
b $q^2 - 169$
e $4x^2 - y^2$
b $16p^2 - 36q^2$
e $(pq)^2 - 4p^2$
b $\frac{1}{4}x^2 - \frac{1}{9}y^2$
e $16x^4 - 81y^4$

c $x^2 - 25$
f $100 - y^2$
c $m^2 - 1$
f $25p^2 - 64q^2$
c $64x^2 - y^2$
f $(ab)^2 - (cd)^2$
c $p^4 - q^4$
f $(2x)^2 - (3y)^4$

Evaluation

Once factorised, numerical expressions can usually be evaluated (worked out) more easily.

Example   Evaluate $13^2 - 7^2$.

$$13^2 - 7^2$$
$$= (13 + 7)(13 - 7)$$
$$= 20 \times 6$$
$$= 120$$

Exercise 5.6

By factorising, evaluate the following.

1 a $8^2 - 2^2$
  d $17^2 - 3^2$

b $16^2 - 4^2$
e $88^2 - 12^2$

c $49^2 - 1$
f $96^2 - 4^2$

2 a $45^2 - 25$ **b** $99^2 - 1$ **c** $27^2 - 23^2$
  d $66^2 - 34^2$ **e** $999^2 - 1$ **f** $225 - 8^2$

3 a $8.4^2 - 1.6^2$ **b** $9.3^2 - 0.7^2$ **c** $42.8^2 - 7.2^2$
  d $8\frac{1}{2}^2 - 1\frac{1}{2}^2$ **e** $7\frac{3}{4}^2 - 2\frac{1}{4}^2$ **f** $5.25^2 - 4.75^2$

4 a $8.62^2 - 1.38^2$ **b** $0.9^2 - 0.1^2$ **c** $3^4 - 2^4$
  d $2^4 - 1$ **e** $1111^2 - 111^2$ **f** $2^8 - 25$

Factorising quadratic expressions

$x^2 + 5x + 6$ is known as a **quadratic expression** because the highest power of any of its terms is squared – in this case x^2. It can be factorised by writing it as a product of two brackets.

Examples   Factorise $x^2 + 5x + 6$.

Set up a 2×2 multiplication grid to help find the factors.

	x	
x	x^2	
		$+6$

As there is only one term in x^2, this can be entered, as can the constant $+6$. The only two values which multiply to give x^2 are x and x. These too can be entered.

We now need to find two values which multiply to give $+6$ and which add to give $+5x$.

	x	$+3$
x	x^2	$3x$
$+2$	$2x$	$+6$

The only two values which satisfy both these conditions are $+3$ and $+2$.
    Therefore $x^2 + 5x + 6 = (x + 3)(x + 2)$.

Factorise $x^2 + 2x - 24$.

	x	
x	x^2	
		-24

	x	$+6$
x	x^2	$+6x$
-4	$-4x$	-24

Therefore $x^2 + 2x - 24 = (x + 6)(x - 4)$.

Exercise 5.7

Factorise the following quadratic expressions.

1 a $x^2 + 7x + 12$ **b** $x^2 + 8x + 12$ **c** $x^2 + 13x + 12$
  d $x^2 - 7x + 12$ **e** $x^2 - 8x + 12$ **f** $x^2 - 13x + 12$
2 a $x^2 + 6x + 5$ **b** $x^2 + 6x + 8$ **c** $x^2 + 6x + 9$
  d $x^2 + 10x + 25$ **e** $x^2 + 22x + 121$ **f** $x^2 - 13x + 42$
3 a $x^2 + 14x + 24$ **b** $x^2 + 11x + 24$ **c** $x^2 - 10x + 24$
  d $x^2 + 15x + 36$ **e** $x^2 + 20x + 36$ **f** $x^2 - 12x + 36$
4 a $x^2 + 2x - 15$ **b** $x^2 - 2x - 15$ **c** $x^2 + x - 12$
  d $x^2 - x - 12$ **e** $x^2 + 4x - 12$ **f** $x^2 - 15x + 36$
5 a $x^2 - 2x - 8$ **b** $x^2 - x - 20$ **c** $x^2 + x - 30$
  d $x^2 - x - 42$ **e** $x^2 - 2x - 63$ **f** $x^2 + 3x - 54$

Algebraic fractions

Sometimes an expression which seems complex can be simplified by factorising first and then cancelling.

Example   Simplify the following algebraic fraction.

$$\frac{x^2 - 3x}{x^2 - 8x + 15}$$

Take each part of the fraction separately and factorise:

$$x^2 - 3x = x(x - 3)$$
$$x^2 - 8x + 15 = (x - 5)(x - 3)$$

Rewrite the original fraction in its factorised form:

$$\frac{x(x - 3)}{(x - 5)(x - 3)}$$

This fraction can be rewritten in the following way:

$$\frac{x}{(x - 5)} \times \frac{\cancel{(x - 3)}}{\cancel{(x - 3)}}$$

Therefore the fraction simplifies to $\dfrac{x}{x - 5}$.

Exercise 5.8

Simplify the algebraic fractions in the following expressions.

1 a $\dfrac{x(x - 4)}{(x - 4)(x + 2)}$ **b** $\dfrac{y(y - 3)}{(y + 3)(y - 3)}$ **c** $\dfrac{(m + 2)(m - 2)}{(m - 2)(m - 3)}$

  d $\dfrac{p(p + 5)}{(p - 5)(p + 5)}$ **e** $\dfrac{m(2m + 3)}{(m + 4)(2m + 3)}$ **f** $\dfrac{(m + 1)(m - 1)}{(m + 2)(m - 1)}$

2 a $\dfrac{x^2 - 5x}{(x+3)(x-5)}$ **b** $\dfrac{x^2 - 3x}{(x+4)(x-3)}$ **c** $\dfrac{y^2 - 7y}{(y-7)(y-1)}$

d $\dfrac{x(x-1)}{x^2 + 2x - 3}$ **e** $\dfrac{x(x+2)}{x^2 + 4x + 4}$ **f** $\dfrac{x(x+4)}{x^2 + 5x + 4}$

3 a $\dfrac{x^2 - x}{x^2 - 1}$ **b** $\dfrac{x^2 + 2x}{x^2 + 5x + 6}$ **c** $\dfrac{x^2 + 4x}{x^2 + x - 12}$

d $\dfrac{x^2 - 5x}{x^2 - 3x - 10}$ **e** $\dfrac{x^2 + 3x}{x^2 - 9}$ **f** $\dfrac{x^2 - 7x}{x^2 - 49}$

> The graph of a quadratic equation $x^2 + bx + c = 0$ is always this shape:
>

Solving quadratic equations

A quadratic equation is an equation where the highest power is 'squared' (for example, x^2). Therefore $x^2 - 5x + 6 = 0$ is an example of a quadratic equation.

When asked to solve a quadratic equation we have to find a value for x (or any other variable) which makes the left-hand side of the equation equal to the right-hand side.

For example, to solve the equation $x^2 - 5x + 6 = 0$ we are being asked to find a value of x which makes $x^2 - 5x + 6$ equal to zero.

One method of solving quadratic equations is to factorise it.

Examples

> It doesn't matter which comes first: $(x-3)(x-2)$ is the same as $(x-2)(x-3)$.

Solve the following quadratic equation: $x^2 - 5x + 6 = 0$.

Factorising $x^2 - 5x + 6$ gives $(x-3)(x-2)$. So $x^2 - 5x + 6 = 0$ becomes

$$(x-3)(x-2) = 0$$

If two numbers multiplied together give zero, then at least one of them must be zero. For $(x-3)(x-2) = 0$, the two numbers are $(x-3)$ and $(x-2)$. Therefore

$$\text{either } (x-3) = 0 \qquad \text{so } x = 3$$
$$\text{or } \quad (x-2) = 0 \qquad \text{so } x = 2$$

We can check the solutions by substituting them into the original equation:
Substituting $x = 3$ into $x^2 - 5x + 6 = 0$ gives:

$$3^2 - (5 \times 3) + 6 = 0$$
$$9 - 15 + 6 = 0$$
$$0 = 0 \qquad \text{a verification}$$

Substituting $x = 2$ into $x^2 - 5x + 6 = 0$ gives:

$$2^2 - (5 \times 2) + 6 = 0$$
$$4 - 10 + 6 = 0$$
$$0 = 0 \qquad \text{a verification}$$

Both solutions are verified.

Solve the quadratic $x^2 + 2x = 24$.

Note. To factorise the quadratic it will need to be arranged so that all the terms are on the same side of the equation.

The equation becomes

$$x^2 + 2x - 24 = 0$$

Factorising gives

$$(x + 6)(x - 4) = 0$$

So either

$$x + 6 = 0 \qquad \text{or} \qquad x - 4 = 0$$

i.e.

$$x = -6 \qquad \text{or} \qquad x = 4$$

Check:

Substituting $x = -6$ into $x^2 + 2x - 24 = 0$ gives:

$$(-6)^2 + (2 \times -6) - 24 = 0$$
$$36 - 12 - 24 = 0$$
$$0 = 0$$

> **Remember:**
> *Always check your solutions by substituting.*

Substituting $x = 4$ into $x^2 + 2x - 24 = 0$ gives:

$$4^2 + (2 \times 4) - 24 = 0$$
$$16 + 8 - 24 = 0$$
$$0 = 0$$

Exercise 5.9

Solve the following quadratic equations by factorising.

1 a $x^2 + 7x + 12 = 0$ **b** $x^2 + 8x + 12 = 0$ **c** $x^2 + 13x + 12 = 0$ **d** $x^2 - 7x + 10 = 0$
2 a $x^2 + 3x - 10 = 0$ **b** $x^2 - 3x - 10 = 0$ **c** $x^2 + 5x - 14 = 0$ **d** $x^2 - 5x - 14 = 0$
3 a $x^2 + 5x = -6$ **b** $x^2 + 6x = -9$ **c** $x^2 + 11x = -24$ **d** $x^2 - 10x = -24$
4 a $x^2 - 2x = 8$ **b** $x^2 - x = 20$ **c** $x^2 + x = 30$ **d** $x^2 - x = 42$

Exercise 5.10

Solve the following quadratic equations.

1 a $x^2 - 9 = 0$ **b** $x^2 - 16 = 0$ **c** $x^2 = 25$ **d** $x^2 = 121$
2 a $4x^2 - 25 = 0$ **b** $9x^2 - 36 = 0$ **c** $25x^2 = 64$ **d** $x^2 = \frac{1}{4}$
3 a $x^2 + 5x + 4 = 0$ **b** $x^2 + 7x + 10 = 0$ **c** $x^2 + 6x + 8 = 0$ **d** $x^2 - 6x + 8 = 0$
4 a $x^2 - 3x - 10 = 0$ **b** $x^2 + 3x - 10 = 0$ **c** $x^2 - 3x - 18 = 0$ **d** $x^2 + 3x - 18 = 0$
5 a $x^2 + x = 12$ **b** $x^2 + 8x = -12$ **c** $x^2 + 5x = 36$ **d** $x^2 + 2x = -1$
6 a $x^2 - 8x = 0$ **b** $x^2 - 7x = 0$ **c** $x^2 + 3x = 0$ **d** $x^2 + 4x = 0$

Exercise 5.11

In the following questions, construct quadratic equations from the information given and then solve to find the unknown.

1 If the area of this rectangle is $10\,\text{cm}^2$, calculate the only possible value of x.

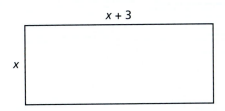

2 If the area of this rectangle is $52\,\text{cm}^2$, calculate the only possible value of x.

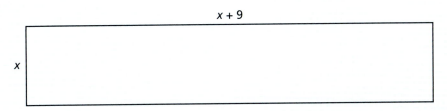

3 A triangle has a base length of $2x\,\text{cm}$ and a height of $(x-3)\,\text{cm}$. If its area is $18\,\text{cm}^2$, calculate its height and base length.

4 A triangle has a base length of $(x-8)\,\text{cm}$ and a height of $2x\,\text{cm}$. If its area is $20\,\text{cm}^2$, calculate its height and base length.

5 A right-angled triangle has a base length of $x\,\text{cm}$ and a height of $(x-1)\,\text{cm}$. If its area is $15\,\text{cm}^2$, calculate the base length and height.

6 A rectangular garden has a square flower bed of side length $x\,\text{m}$ in one of its corners. The remainder of the garden consists of lawn and has dimensions as shown. The total area of the lawn is $50\,\text{m}^2$.

  a Form an equation in terms of x.

  b Solve the equation.

  c Calculate the length and width of the whole garden.

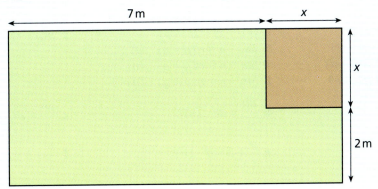

Transformation of formulae

In *Intermediate 1* you learned how to rearrange a formula to change the subject. For example, in the formula $a = 2b + c$, a is the subject. To make either b or c the subject, the formula has to be rearranged. To rearrange a formula, there is one law that must be obeyed every time, namely:

● *What is done to one side of the formula must also be done to the other side.*

Example Rearrange the following formulae to make the letter in brackets the subject.

a $a = 2b + c$ (c)
b $2r + p = q$ (p)
c $ab = cd$ (c)
d $\dfrac{a}{b} = \dfrac{c}{d}$ (d)

> **Remember:**
> This is called **changing the subject** of the formula.

a $a = 2b + c$
$a - 2b = c$

b $2r + p = q$
$p = q - 2r$

c $ab = cd$
$\dfrac{ab}{d} = c$

d $\dfrac{a}{b} = \dfrac{c}{d}$
$ad = cb$
$d = \dfrac{cb}{a}$

Exercise 5.12

In the following questions, make the letter in brackets at the end the subject of the formula.

1 a $m + n = r$ (n)
b $m + n = p$ (m)
c $2m + n = 3p$ (n)
d $3x = 2p + q$ (q)
e $ab = cd$ (a)
f $ab = cd$ (d)

2 a $3xy = 4m$ (x)
b $7pq = 5r$ (r)
c $3x = c$ (x)
d $3x + 7 = y$ (x)
e $5y - 9 = 3r$ (y)
f $5y - 9 = 3x$ (x)

3 a $6b = 2a - 5$ (b)
b $6b = 2a - 5$ (a)
c $3x - 7y = 4z$ (z)
d $3x - 7y = 4z$ (x)
e $3x - 7y = 4z$ (y)
f $2pr - q = 8$ (p)

4 a $\dfrac{p}{4} = r$ (p)
b $\dfrac{4}{p} = 3r$ (p)
c $\dfrac{1}{5}n = 2p$ (p)

d $\frac{1}{5}n = 2p$ (n)
e $p(q + r) = 2t$ (p)
f $p(q + r) = 2t$ (q)

5 a $3m - n = rt(p + q)$ (r)
b $3m - n = rt(p + q)$ (t)
c $3m - n = rt(p + q)$ (m)
d $3m - n = rt(p + q)$ (n)
e $3m - n = rt(p + q)$ (p)
f $3m - n = rt(p + q)$ (q)

6 a $\dfrac{ab}{c} = de$ (d)
b $\dfrac{ab}{c} = de$ (a)
c $\dfrac{ab}{c} = de$ (c)

d $\dfrac{a + b}{c} = d$ (a)
e $\dfrac{a}{c} + b = d$ (b)
f $\dfrac{a}{c} + b = d$ (c)

More complex formulae obey the same basic rule when they are transformed, i.e. what is done to one side of the formula must also be done to the other. If you remember this, you will find it is also possible to rearrange formulae involving squares and square roots as shown below.

Example Make the letter in brackets the subject of each formula.

a $C = 2\pi r$ (r) **b** $A = \pi r^2$ (r)
c $Rx^2 = p$ (x) **d** $x^2 + y^2 = h^2$ (y)

e $\sqrt{x} = tv$ (x) **f** $f = \sqrt{\dfrac{x}{k}}$ (x)

g $m = 3a\sqrt{\dfrac{p}{x}}$ (x) **h** $x + a = \dfrac{x+y}{p}$ (x)

Remember:

$\dfrac{C}{2\pi} = r$ is the same as

$r = \dfrac{C}{2\pi}$

a $C = 2\pi r$

$$\frac{C}{2\pi} = r$$

b $A = \pi r^2$

$$\frac{A}{\pi} = r^2$$

$$\sqrt{\frac{A}{\pi}} = r$$

c $Rx^2 = p$

$$x^2 = \frac{p}{R}$$

$$x = \sqrt{\frac{p}{R}}$$

d $x^2 + y^2 = h^2$

$$y^2 = h^2 - x^2$$

$$y = \sqrt{h^2 - x^2} \quad \text{[Note: } not$$
$$y = h - x]$$

e $\sqrt{x} = tv$

$$x = t^2 v^2$$

$$\text{or } x = (tv)^2$$

f $f = \sqrt{\dfrac{x}{k}}$

$$f^2 = \frac{x}{k}$$

$$f^2 k = x$$

g $m = 3a\sqrt{\dfrac{p}{x}}$

$$m^2 = \frac{9a^2 p}{x} \quad \text{(square both sides)}$$

$$m^2 x = 9a^2 p$$

$$x = \frac{9a^2 p}{m^2}$$

h $x + a = \dfrac{x+y}{p}$

$$p(x + a) = x + y$$

$$px + pa = x + y$$

$$px - x = y - pa$$

$$x(p - 1) = y - pa$$

$$x = \frac{y - pa}{p - 1}$$

Exercise 5.13

In each of the formulae below, make x the subject.

1 a $P = 2mx$
b $T = 3x^2$
c $mx^2 = y^2$
d $x^2 + y^2 = p^2 - q^2$
e $m^2 + x^2 = y^2 - n^2$
f $p^2 - q^2 = 4x^2 - y^2$

2 a $\dfrac{P}{Q} = rx$
b $\dfrac{P}{Q} = rx^2$
c $\dfrac{P}{Q} = \dfrac{x^2}{r}$

d $\dfrac{m}{n} = \dfrac{1}{x^2}$
e $\dfrac{r}{st} = \dfrac{w}{x^2}$
f $\dfrac{p+q}{r} = \dfrac{w}{x^2}$

3 a $\sqrt{x} = rp$
b $\dfrac{mn}{p} = \sqrt{x}$
c $g = \sqrt{\dfrac{k}{x}}$

d $pq - px = q + x$
e $p(x - r) = x + r$
f $x + a = \dfrac{x+b}{y}$

SUMMARY

By the time you have completed this chapter you should know:

■ how to solve an equation by **trial and improvement**
■ how to factorise simple expressions, for example
$$10x - 25 = 5(2x - 5)$$

■ how to factorise by grouping, for example
$$6x + 3 + 2xy + y$$
$$= 3(2x + 1) + y(2x + 1)$$
$$= (3 + y)(2x + 1)$$

■ how to factorise by using the **difference of two squares**, for example
$$9x^2 - 4y^2$$
$$= (3x - 2y)(3x + 2y)$$

■ how to factorise **quadratic expressions**, for example
$$x^2 - 5x + 6$$
$$= (x - 3)(x - 2)$$

■ how to simplify by cancelling, for example
$$\frac{x(x-5)}{(x+3)(x-5)} = \frac{x}{x+3}$$

■ how to solve quadratic equations by factorising, for example
$$x^2 + 6x + 8 = 0$$
$$(x + 4)(x + 2) = 0$$
therefore $x = -4$ or -2

■ how to **change the subject** of complex formulae, for example make p the subject of the formula below:

a $p^2 + q^2 = r^2$
$p^2 = r^2 - q^2$
$p = \sqrt{r^2 - q^2}$

b $p + py = mn - y$
$p(1 + y) = mn - y$
$p = \dfrac{mn - y}{1 + y}$

Exercise 5A

1 Expand the following and simplify where possible.
   a $(x+3)(x+5)$ **b** $(x-7)(x-7)$
   c $(x+5)^2$ **d** $(2x-1)(3x+8)$

2 Make the letter in red the subject of the formula.
   a $v^2 = u^2 + 2as$ **b** $r^2 + h^2 = p^2$

   c $\dfrac{m}{n} = \dfrac{r}{s^2}$ **d** $T = 2\pi\sqrt{\dfrac{L}{g}}$

3 Expand the following and simplify where possible.
   a $(x-4)(x+2)$ **b** $(x-8)^2$
   c $(x+y)^2$ **d** $(x-11)(x+11)$

4 Make the letter in red the subject of the formula.
   a $mf^2 = p$ **b** $m = 5t^2$

   c $A = \pi r\sqrt{p+q}$ **d** $\dfrac{1}{x} + \dfrac{1}{y} = \dfrac{1}{t}$

5 Factorise the following fully.
   a $mx - 5m - 5nx + 25n$ **b** $4x^2 - 81y^2$ **c** $x^4 - y^4$

6 Factorise the following fully.
   a $pq - 3rq + pr - 3r^2$ **b** $1 - t^4$

Exercise 5B

1 Factorise the following.
   a $x^2 - 18x + 32$ **b** $x^2 - 2x - 24$
   c $x^2 - 9x + 18$ **d** $x^2 - 2x + 1$

2 Solve the following quadratic equations.
   a $x^2 + 5x = -6$ **b** $x^2 = 4x - 4$

3 Simplify the following algebraic expressions.

   a $\dfrac{x(x-2)}{(x-1)(x-2)}$ **b** $\dfrac{x^2 + 3x}{x^2 + x - 6}$

4 Factorise the following expressions.
   a $x^2 - 4x - 77$ **b** $x^2 - 6x + 9$ **c** $x^2 - 144$

5 Solve the following quadratic equations.
   a $x^2 = 9$ **b** $6x^2 = 150$

6 Simplify the following algebraic expressions.

   a $\dfrac{(m+4)(m-3)}{(m-3)}$ **b** $\dfrac{m^2 + 4m}{m^2 - 16}$

Exercise 5C Ma1

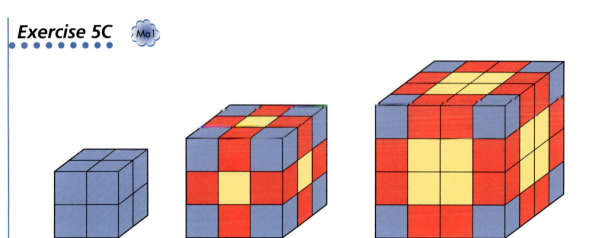

The cubes above are made from smaller 1×1 cubes. They are painted according to the following rules:

1×1 cubes with 3 faces showing are painted blue.
1×1 cubes with 2 faces showing are painted red.
1×1 cubes with 1 face showing are painted yellow.
1×1 cubes with 0 faces showing are painted white.

1 For each of the cubes shown above, count the number of blue, red, yellow and white 1×1 cubes.
2 Predict how many of each colour of cube there will be in a larger 5×5 cube.
3 For each colour of cube, write an algebraic formula for the number of that colour of cubes to be found in an $N \times N$ cube.

Exercise 5D Ma1

You will need:
• computer with spreadsheet package installed

Set up a spreadsheet to display the results of the problem in exercise 5C.
It may look similar to the one shown below:

Cube side length	Number of blue cubes	Number of red cubes	Number of yellow cubes	Number of white cubes
2				
3				
4				
5				

Use formulae to generate the numbers in these cells

Use your spreadsheet to predict the number of each colour of cube present in a large 10×10 cube.

Exercise 5E Ma1

You will need:
• computer with internet access

Michael Faraday (1791–1867) was a brilliant experimental scientist who invented, among other things, the electric motor. James Maxwell (1831–1879) was a Scottish physicist and mathematician.

Using the internet as a resource, see if you can find how Maxwell's work was linked to Faraday's experiments.

6 : Graphs of functions

In *Intermediate 1*, you learned how to plot a graph from a **linear** equation, for example $y = 2x + 3$. In this chapter you will plot graphs that are straight lines, and also graphs that are curved. If the graph of $y = x^2$ were to be plotted, its shape would form a special type of curve known as a **parabola**.

A **helix** (or spiral) has some similarity to a parabola, but it curves through three dimensions, while a parabola is two-dimensional.

The most famous example of a double helix is the structure of the DNA molecule. The discovery of DNA was made by geneticists Francis Crick and James Watson in 1953. Their discovery was made possible by the study of X-ray pictures made by Rosalind Franklin. These pictures gave them information on the structure of proteins that enabled Watson and Crick to build a model of the DNA molecule like the one shown.

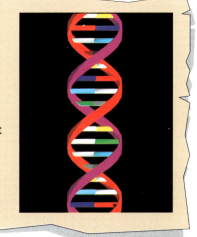

Linear graphs

> The point where the line crosses the y-axis is the y-intercept.

In *Intermediate 1* we saw that the equation of a straight-line graph can be written in the form $y = mx + c$, where m represents the gradient (slope) of the straight line and c represents where the line intersects the y-axis.

A summary is shown below:

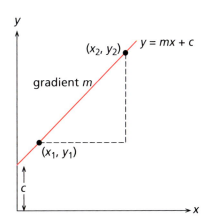

The gradient is calculated by taking two points on the line, working out the vertical distance between them $(y_2 - y_1)$ and dividing this by the horizontal distance between them, i.e. $(x_2 - x_1)$. Therefore

$$m = \frac{y_2 - y_1}{x_2 - x_1}$$

Remember:

$$Gradient = \frac{y\text{-}change}{x\text{-}change}$$

Note. Positive gradients go uphill from left to right ╱

Negative gradients go downhill from left to right ╲

Exercise 6.1

(*Revision*)

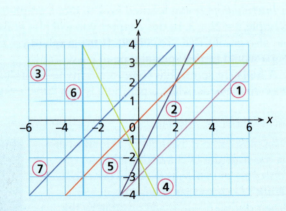

Using the graph above, answer the following questions.

1 Which of the lines 1–7 have the following equations?
  a $y = x + 2$ **b** $y = 3$ **c** $x = -3$
  d $y = 2x - 2$ **e** $y = x$ **f** $y = x - 3$
  g $y = -2x - 2$
2 By looking at the equations, how can you decide which lines are parallel? Explain your answer fully.
3 Explain why the line $y = 2x - 2$ has a gradient double that of line $y = x + 2$.

Exercise 6.2

(*Revision*)
1 By drawing axes to help if necessary, calculate the gradient of the line passing through each of the following pairs of points:
  a $(1, 3)$ and $(3, 7)$ **b** $(0, 6)$ and $(1, 11)$
  c $(-2, 4)$ and $(1, 13)$ **d** $(7, 5)$ and $(9, 1)$
2 On a suitable grid with a pair of axes, draw the following straight lines.
  a A line with a gradient of 2 passing through the point $(0, 2)$
  b A line with a gradient of -1 passing through the point $(-1, -1)$.

Solving simultaneous equations graphically

In chapter 5 we looked at simultaneous equations and how they could be solved using a method of elimination. For example, solve the following equations simultaneously by eliminating either x or y:

$$2x - y = 4 \qquad (1)$$
$$x + y = 8 \qquad (2)$$

Adding equations (1) and (2) removes y:

$$\begin{array}{r} 2x - y = 4 \\ x + y = 8 \\ \hline 3x \quad\;\; = 12 \end{array}$$

Therefore $x = 4$.

Substituting $x = 4$ into equation (2) gives:

$$4 + y = 8$$

Therefore $y = 4$.

So the solution $x = 4$ and $y = 4$ satisfies both equations.

If both equations are plotted on the same diagram we get the graph on the right.

From the diagram it can be seen that the point where the two graphs intersect is (4, 4), i.e. $x = 4$ and $y = 4$. Therefore, when two equations are plotted simultaneously, their point of intersection represents the solution of the two simultaneous equations, i.e. it is the *only* point that lies on *both* lines.

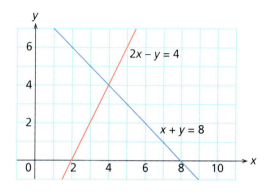

Exercise 6.3

1 Solve the following pairs of simultaneous equations by plotting them on a suitable grid and finding their point of intersection.

a $x + y = 5$
$x - y = 1$

b $x + y = 7$
$x - y = 3$

c $2x + y = 5$
$x - y = 1$

d $2x + 2y = 6$
$2x - y = 3$

e $x + 3y = -1$
$x - 2y = -6$

f $x - y = 6$
$x + y = 2$

2 Solve the following simultaneous equations by plotting them on a suitable grid and finding their point of intersection.

a $3x - 2y = 13$
$2x + y = 4$

b $4x - 5y = 1$
$2x + y = -3$

c $x + 5 = y$
$2x + 3y - 5 = 0$

d $x = y$
$x + y + 6 = 0$

3 Entrance to a theatre costs each adult £x and each child £y. Three adults and five children pay £25, while four adults and two children pay £24.

a Write down two equations that represent the information given above.

b By plotting both equations on a suitable grid with axes, find the values of x and y.

4 A father and daughter are aged F years old and D years old respectively. At present the father is three times older than his daughter. In 15 years' time the father will be twice as old as his daughter.

a Write down two equations that represent the information given above.

b By plotting both equations on a suitable grid with axes, find the values of F and D.

5

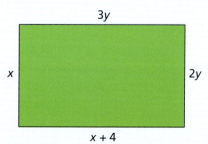

a Write down two equations that represent the information given above.

b By plotting both equations on a suitable grid with axes, find the values of x and y.

c Check your answers by also solving the equations simultaneously.

Quadratic functions

So far we have looked at equations in the form $y = mx + c$. However, equations linking x and y can take other forms, for example:

$$y = x^2 \qquad y = \tfrac{1}{2}x^2 + 3 \qquad y = 3x^2 + x + 1$$

All the above equations are of the form $y = ax^2 + bx + c$. As the highest power of x is x^2, these are known as **quadratic equations** (or quadratic functions). When plotted, they do not make a straight line.

For example, plot the graph of $y = x^2$, where x lies in the range $-4 \leqslant x \leqslant 4$. A table of results gives the coordinates for the plotted points.

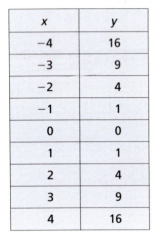

x	y
−4	16
−3	9
−2	4
−1	1
0	0
1	1
2	4
3	9
4	16

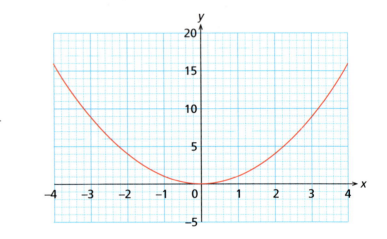

Example   Plot a graph of the function $y = -x^2 + x + 2$ for the range $-3 \leqslant x \leqslant 4$.

Draw up a table of values:

x	−3	−2	−1	0	1	2	3	4
y	−10	−4	0	2	2	0	−4	10

The graph of the function is given below:

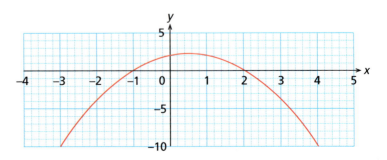

Exercise 6.4

For each of the following quadratic functions, construct a table of values for the range indicated and then draw the graph.

1 $y = x^2 + x - 2$ $-4 \leqslant x \leqslant 3$
2 $y = -x^2 + 2x + 3$ $-3 \leqslant x \leqslant 5$
3 $y = x^2 - 4x + 4$ $-1 \leqslant x \leqslant 5$
4 $y = -2x^2 + x + 6$ $-3 \leqslant x \leqslant 3$
5 $y = 4x^2 + 4x - 1$ $-3 \leqslant x \leqslant 2$

> Be careful when you choose scales for your axes!

Graphical solutions of quadratic equations

Just as graphing linear equations helps us solve them, graphing quadratics can also be helpful when it comes to solving them.

Example   Draw a graph of $y = x^2 - 4x + 3$ for the range $-2 \leqslant x \leqslant 5$. Use the graph to solve the equation $x^2 - 4x + 3 = 0$.

The table of values and graph are as follows:

x	y
-2	15
-1	8
0	3
1	0
2	-1
3	0
4	3
5	8

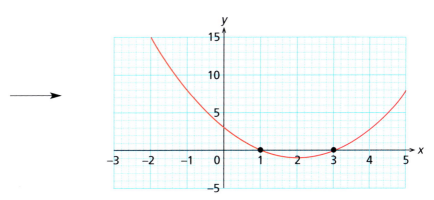

To solve the equation we need to find the values of x when $y = 0$, i.e. where the graph crosses the x-axis. As can be seen from the graph above, these points occur when $x = 1$ and $x = 3$.

So $x = 1$ and $x = 3$ are the solutions of the equation $x^2 - 4x + 3 = 0$.

Exercise 6.5

Solve each of the quadratic functions below by plotting a graph for the ranges of x given. Keep these graphs – you will need them for exercise 6.6.

1 $x^2 - x - 6 = 0$ $-4 \leqslant x \leqslant 4$ 2 $-x^2 + 1 = 0$ $-4 \leqslant x \leqslant 4$
3 $x^2 - 6x + 9 = 0$ $0 \leqslant x \leqslant 6$ 4 $-x^2 - x + 12 = 0$ $-5 \leqslant x \leqslant 4$
5 $x^2 - 4x + 4 = 0$ $-2 \leqslant x \leqslant 6$ 6 $2x^2 - 7x + 3 = 0$ $-1 \leqslant x \leqslant 5$
7 $-2x^2 + 4x - 2 = 0$ $-2 \leqslant x \leqslant 4$ 8 $3x^2 - 5x - 2 = 0$ $-1 \leqslant x \leqslant 3$

In the previous worked example, as $y = x^2 - 4x + 3$, a solution could be found to the equation $x^2 - 4x + 3 = 0$ by reading off the points where the graph crossed the x-axis. The graph can, however, also be used to solve other quadratic equations.

Example   Use the graph of $y = x^2 - 4x + 3$ to solve the equation $x^2 - 4x + 1 = 0$.

$x^2 - 4x + 1 = 0$ can be rearranged as follows:

$$x^2 - 4x + 3 = 2 \qquad \text{(adding 2 to both sides)}$$

Draw the graph of $y = x^2 - 4x + 3$ and plot the line $y = 2$ on the same axes to give the graph shown below:

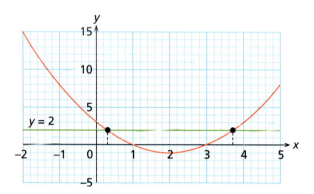

The points where the curve and the line intersect give the solutions to $x^2 - 4x + 3 = 2$ and hence also $x^2 - 4x + 1 = 0$.
From the graph we can see that the solutions to $x^2 - 4x + 3 = 2$ are $x \approx 0.3$ and $x \approx 3.7$. Note that, because of the scale used for the graph, these solutions are only approximate ($\approx$ means 'is approximately equal to').

Exercise 6.6

Using the graphs that you drew in exercise 6.5, solve the following quadratic equations. Show your method clearly.

1 $x^2 - x - 4 = 0$
2 $-x^2 - 1 = 0$
3 $x^2 - 6x + 8 = 0$
4 $-x^2 - x + 9 = 0$
5 $x^2 - 4x + 1 = 0$
6 $2x^2 - 7x = 0$
7 $-2x^2 + 4x = -1$
8 $3x^2 = 2 + 5x$

You will need:
- the graphs you drew for exercise 6.5

The reciprocal function

We have seen that quadratic functions produce a distinctively shaped curve in the form of a parabola. Functions of the form $y = \dfrac{a}{x}$ (i.e. where x is the denominator) also produce a distinctively shaped curve.

Example

Draw the graph of $y = \dfrac{2}{x}$ for $-4 \leqslant x \leqslant 4$.

x	-4	-3	-2	-1	0	1	2	3	4
y	-0.5	-0.7	-1	-2	$-$	2	1	0.7	0.5

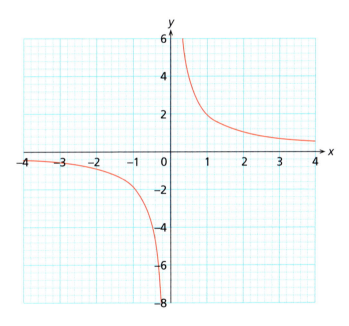

This is a **reciprocal function** and the shape of the graph is known as a **hyperbola**.

Note. When $x = 0$, there is no possible value of y, therefore there is a 'gap' in the graph.

Exercise 6.7

1 Plot the graph of the function $y = \dfrac{1}{x}$ for $-4 \leqslant x \leqslant 4$.

2 Plot the graph of the function $y = \dfrac{3}{x}$ for $-4 \leqslant x \leqslant 4$.

3 a Plot the graph of the function $y = \dfrac{5}{3x}$ for $-4 \leqslant x \leqslant 4$.

   b Use your graph to find an approximate solution to the equation $\dfrac{5}{3x} + 3 = 0$.

Exercise 6.8

By looking at the shape of each of the graphs below, decide whether they show a linear, quadratic or reciprocal function.

1

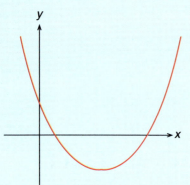

2

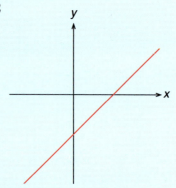

3

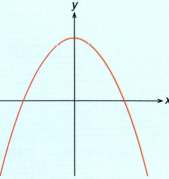

4

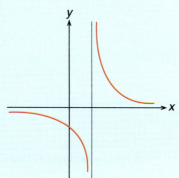

5

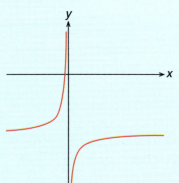

6

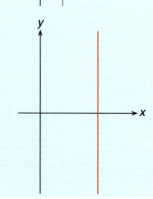

The cubic function

We have seen that functions where the highest power is squared (for example, x^2) are called quadratic functions and their shape when plotted is called a parabola. Functions where the highest power is cubed (for example x^3) are called **cubic functions**. Their shape when plotted is known as a **cubic curve**. It is easy to recognise by its S-like shape.

Example

a Plot the graph of $y = 2x^3 + 10x^2 + x - 10$ for x in the range $-5 \leqslant x \leqslant 2$.

b Use the graph to find an approximate solution to $2x^3 + 10x^2 + x - 10 = 0$.

a For complex equations such as this, it is often easier to construct a table that is broken down into steps.

x
-5
-4
-3
-2
-1
0
1
2

$\longrightarrow$

$2x^3$	$10x^2$	x	-10
-250	250	-5	-10
-128	160	-4	-10
-54	90	-3	-10
-16	40	-2	-10
-2	10	-1	-10
0	0	0	-10
2	10	1	-10
16	40	2	-10

$\longrightarrow$

y
-15
-18
23
12
-3
10
3
48

The values of x and y can then be plotted to give the graph below.

> Because of the scale used for the graph, these solutions are only approximate.

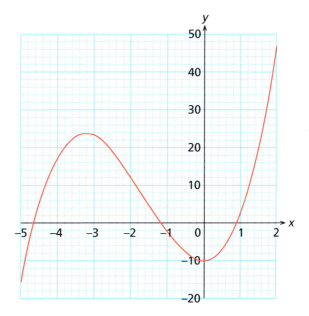

b The solution to $2x^3 + 10x^2 + x - 10 = 0$ is found when $y = 0$, i.e. where the graph crosses the x-axis. From the graph, we can see this happens three times: $x \approx 0.8$, -1.2 and -4.7.

Exercise 6.9

1 **a** Plot the graph of $y = x^3$ in the range $-3 \leqslant x \leqslant 3$.
  b Use your graph to find the solution to $x^3 = 0$.
2 **a** Plot the graph of $y = \frac{1}{2}x^3 - x$ in the range $-3 \leqslant x \leqslant 3$.
  b Use your graph to find the solution to $\frac{1}{2}x^3 - x = 0$.
3 **a** Plot the graph of $y = (x^2 - 4)(x - 2)$ in the range $-3 \leqslant x \leqslant 3$.
  b Use your graph to find the solution to $(x^2 - 4)(x - 2) = 0$.
4 Use your graphs drawn for questions 1–3, and graphs of any other cubics of your choice, to make a conclusion about the general shape of a cubic curve.
5 What conclusions can you make about the number of solutions possible for a cubic function that is equal to zero?

SUMMARY

By the end of this chapter you should know:

■ that a **linear function** is represented by the equation $y = mx + c$ and that m represents the gradient of the straight line, while c represents the intercept with the y-axis.

■ that when two lines are plotted on a grid with the same axes, their point of intersection represents the solution to the two simultaneous equations

■ that, when a **quadratic function** is plotted, the graph has the shape of a **parabola**

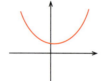
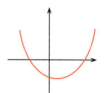

■ that the graph of one **quadratic function** can be used to solve another quadratic function, for example the graph of $y = x^2$ can be used to solve the equation $x^2 = 4$, if the line $y = 4$ is plotted on the same axes

■ that the shape of the graph of a **reciprocal function** is known as a **hyperbola**

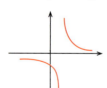

■ that the graph of a **cubic function** is known as a **cubic curve** and has an S-like shape

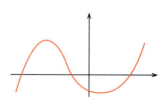

■ how to distinguish the type of function by the shape of its graph.

Exercise 6A

1 *Sketch* the graph of the function $y = \dfrac{1}{x}$.

2 a Copy and complete the table below for the function $y = -x^2 - 7x - 12$.

x	−7	−6	−5	−4	−3	−2	−1	0
y		−6				−2		

   b Plot a graph of your results from the table above.
   c Use your graph to give solutions to the equation $-x^2 - 7x - 12 = 0$.

3 a Plot the graph of the quadratic equation $y = -x^2 - x + 15$ for $-6 \leqslant x \leqslant 4$.
   b Showing your method clearly, use your graph to solve the equation $x^2 = x + 5$.

4 Identify the type of function that produces each of the graphs below.

a

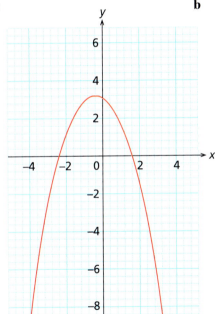

b

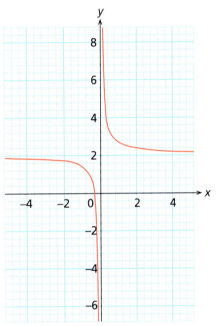

Exercise 6B

1 *Sketch* the graph of each of the following functions.
   a $y = x^2$
   b $y = -x^2$

2 a Copy and complete the table below for the function $y = x^2 + 8x + 15$.

x	−7	−6	−5	−4	−3	−2	−1	0	1	2
y		3				3				

   b Plot a graph of the function.

3 a Plot a graph of the quadratic function $y = x^2 + 9x + 20$ for the range $-7 \leqslant x \leqslant 0$.
   b Showing your method clearly, use your graph to solve the equation $x^2 = -9x - 14$.

4 a Copy and complete the table below for the function $y = -x^3 + 3x + 1$.

x	−2	−1	0	1	2
y					

b Using the results in your table, plot a graph of the function.
c Using your graph, give approximate solutions to the equation $-x^3 + 3x + 1 = 0$.

Exercise 6C

When designing a cereal packet, the manufacturer needs to minimise costs. One of the costs that the manufacturer tries to reduce is the cost of the packaging. In the examples below, two packets A and B have the same volume:

You will need:
• computer with spreadsheet package installed

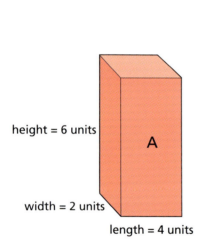

height = 6 units
A
width = 2 units
length = 4 units

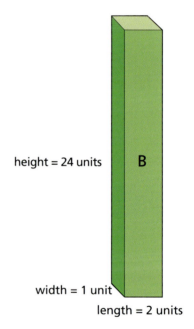

height = 24 units
B
width = 1 unit
length = 2 units

volume of A is $6 \times 2 \times 4 = 48$ units^3
volume of B is $24 \times 1 \times 2 = 48$ units^3

However, the surface area of each packet is very different:

surface area of A is $(2 \times 6 \times 2) + (2 \times 2 \times 4) + (2 \times 4 \times 6) = 88$ units^2
surface area of B is $(2 \times 24 \times 1) + (2 \times 1 \times 2) + (2 \times 24 \times 2) = 148$ units^2

Therefore the cost of producing packet B is greater as it has a greater surface area and uses more card.
   A cereal manufacturer wants to produce a packet with a volume of 5550 cm^3. It must be cuboid in shape and its height must be 30 cm.
   Using a spreadsheet, calculate the smallest surface area that the cereal packet can have. You may wish to set up your spreadsheet in a similar way to the one shown at the top of page 85.

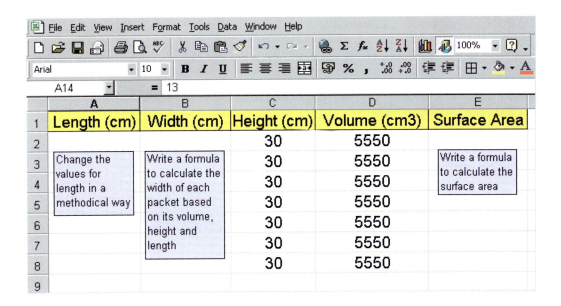

The following spreadsheet content is shown:

	A	B	C	D	E
	Length (cm)	Width (cm)	Height (cm)	Volume (cm3)	Surface Area
2			30	5550	
3	Change the values for length in a methodical way	Write a formula to calculate the width of each packet based on its volume, height and length	30	5550	Write a formula to calculate the surface area
4			30	5550	
5			30	5550	
6			30	5550	
7			30	5550	
8			30	5550	
9					

Cell reference: A14 = 13

Exercise 6D

You will need:
• computer with graphing package installed

Quadratic functions take the general form $y = ax^2 + bx + c$. Each of the variables a, b and c affects the shape of the parabola in some way.

Using the graphing package 'Autograph' or similar, investigate the effect of each variable using the 'constant controller' facility. An example is shown in the screen shot below.

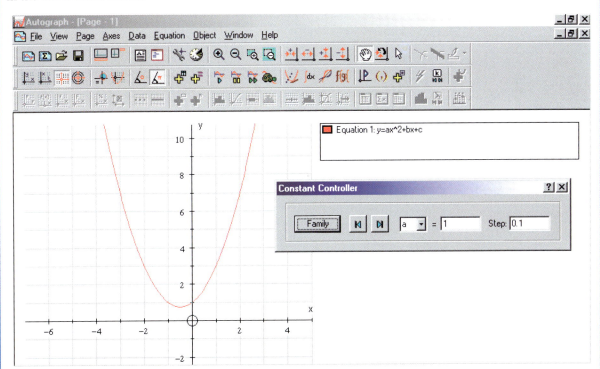

Exercise 6E

You will need:
• computer with internet access

The Nobel Prize was awarded to James Watson and Francis Crick for their work on the structure of DNA.

• Using the internet as a resource, find out who founded the Nobel Prize. Find out also how he made his fortune that enabled him to institute the prize.
• There were originally five Nobel Prizes (later extended to six). What are they awarded for?

7 Constructions and loci

You will know that the atom is composed of a nucleus containing neutrons and protons. Electrons travel in orbits around the nucleus. The orbit of the electrons could be called the **locus** of the electrons.

Maria Goeppert Mayer shared the Nobel Prize for Physics in 1963. She was the second woman to be awarded the Nobel Prize for Physics after Marie Curie. She trained as a mathematician before changing to physics. She investigated the elements with complete shells, the 'noble gases' (helium, neon, argon, krypton, xenon and radon) and explained the stability of the nuclei of these elements.

This diagram shows the atomic structure of krypton. It shows the nucleus at the centre, surrounded by electrons travelling in orbits around it.

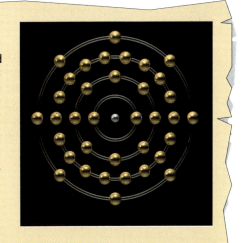

The **locus** (plural **loci**) of a point is the path or orbit it follows as it moves. The main types of loci are explained below.

The locus of the points which are at a given distance from a given point

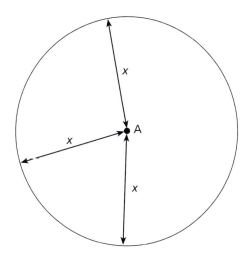

In the diagram above, we can see that the locus of all the points equidistant from a point A lie on the circumference of a circle centre A. This is because all points on the circumference of a circle are equidistant from the centre.

Remember:
Equidistant means 'the same distance'.

The locus of the points which are at a given distance from a given straight line

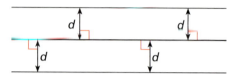

In the diagram above, we can see that the locus of the points equidistant from a straight line runs parallel to that straight line on either side. It is important to note that the distance of the locus from the straight line must be measured at right angles to the line. Note that the line continues to infinity in each direction.

The diagram below shows only part of a line: a line segment. The locus of the points equidistant from a line segment AB takes the form shown. At each end a semi-circle is formed.

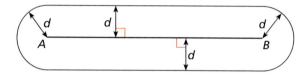

Revision: the perpendicular bisector

A line AB is drawn below. Construct the **perpendicular bisector** of AB.

- Open a pair of compasses to more than half the distance AB.
- Place the compass point on A and draw arcs above and below AB.
- With the compasses kept the same distance apart, place the compass point on B and draw two more arcs above and below AB. Note that the two pairs of arcs should intersect.
- Draw a line through the two points where the arcs intersect.

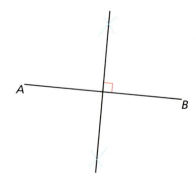

The line drawn is the perpendicular bisector of AB, as it divides AB in half and also meets it at right angles.

The locus of the points which are equidistant from two given points

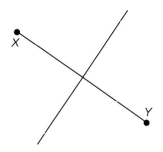

The locus of the points **equidistant** from points X and Y lies on the perpendicular bisector of the line XY.

> **Remember:**
> *A perpendicular bisector is a line which divides another one in half and meets it at right angles.*

Revision: bisecting an angle

Using a pair of compasses, bisect the angle A on the right.

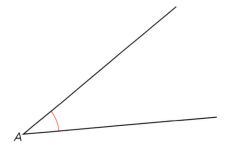

- Open a pair of compasses and place the point on A. Draw two arcs so that they intersect the lines (see right).

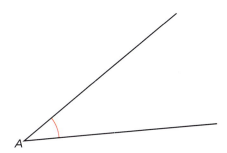

- Place the point of the compasses in turn on the points of intersection and draw another pair of arcs inside the angle. Ensure that they intersect.
- Draw a line through A and the point of intersection of the two arcs. This line bisects angle A (see below).

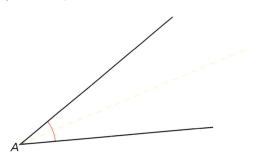

The locus of the points which are equidistant from two given intersecting straight lines

The intersecting straight lines are labelled AB and CD below. The locus in this case lies on the bisectors of both pairs of opposite angles as shown.

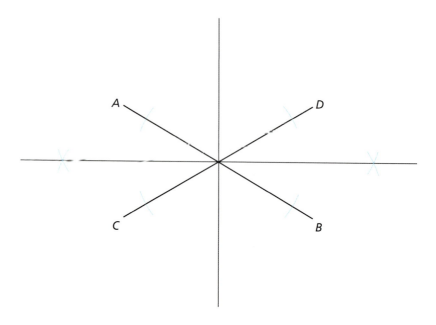

The application of the above cases will enable you to tackle problems involving loci.

Example   The diagram below shows a trapezoidal garden. Three of its sides are enclosed by a fence; the fourth is next to a house.

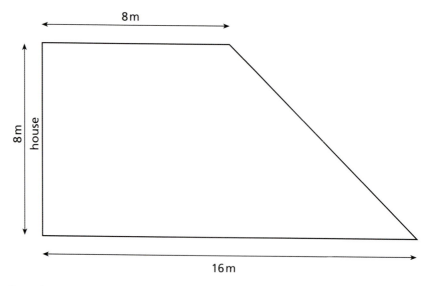

a Grass is to be planted in the garden. However, it must be at least 2 m away from the house and at least 1 m away from the fence. Shade the region in which the grass can be planted.

b Using the same garden as previously, grass must now be planted according to the following conditions: it must be *more than* 2 m away from the house and *more than* 1 m away from the fence. Shade the region in which the grass can be planted.

a The shaded region is all the points which are at least 2 m away from the house and at least 1 m away from the surrounding fence. Note that the boundary of the region is formed by the locus of the points, so it is drawn as a solid line.

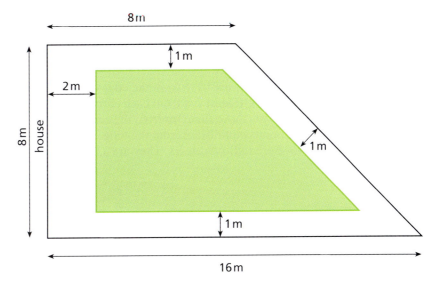

b The shape of the region is the same as in the first case, but in this instance the boundary is *not* included in the locus of the points as the grass cannot be 2 m away from the house or 1 m away from the fence. So the boundary is drawn as a broken line.

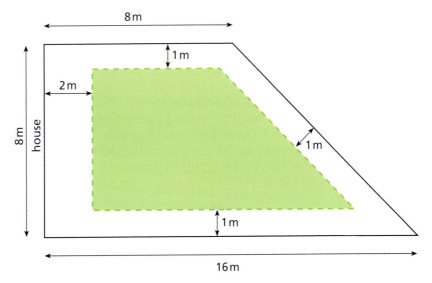

Therefore, if the locus of points is included in the region, it is represented by a *solid* line. If it is not included, then it is represented by a *broken* line.

Exercise 7.1

Questions 1–4 are about a rectangular garden 8 m by 6 m. For each question, draw a scale diagram of the garden and identify the region which fits the criteria.

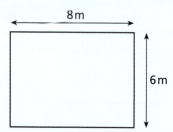

1 Draw the locus of all the points at least 1 m from the edge of the garden.
2 Draw the locus of all the points at least 2 m from each corner of the garden.
3 Draw the locus of all the points more than 3 m from the centre of the garden.
4 Draw the locus of all the points equidistant from the longer sides of the garden.
5 *X* and *Y* are two ship-to-shore radio receivers. They are 25 km apart.

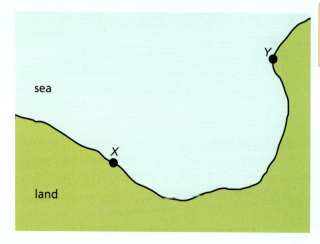

Remember:
*Solid line means **included**, broken line means **not** included.*

A ship sends out a distress signal. The signal is picked up by both *X* and *Y*. The radio receiver at *X* indicates that the ship is within a 30 km radius of *X*, while the radio receiver at *Y* indicates that the ship is within 20 km of *Y*.
Draw a scale diagram and identify the region in which the ship must lie.
6 Draw a line *AB* 8 cm long. What is the locus of a point *C* such that the angle *ACB* is always a right angle?
7 Three lionesses L_1, L_2 and L_3 have surrounded a gazelle. The three lionesses are equidistant from the gazelle. Draw a diagram with the lionesses in similar positions to those shown below and by construction determine the position (*g*) of the gazelle.

L_1 •   •L_2

L_3 •

Exercise 7.2

1 Three girls are playing hide and seek. Ayshe and Belinda are at the positions shown below and are trying to find Cristina. Cristina is on the opposite side of a wall *PQ* to her two friends.

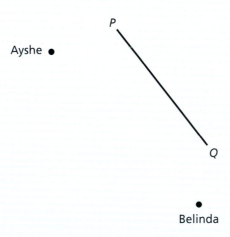

Assuming Ayshe and Belinda cannot see over the wall, identify, by copying the above diagram, the locus of points where Cristina could be if:
a Cristina can only be seen by Ayshe
b Cristina can only be seen by Belinda
c Cristina cannot be seen by either of her two friends
d Cristina can be seen by both of her friends.

2 A security guard *S* is inside a building in the position shown. The building is inside a rectangular compound.

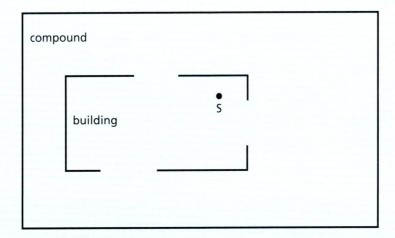

The building has three windows as shown. Identify the locus of points in the compound which can be seen by the security guard.

3 The circular cage shown below houses a snake. Inside the cage are three obstacles.

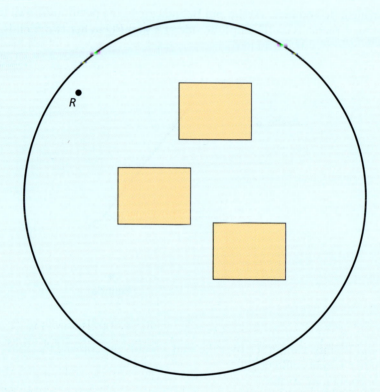

A rodent is placed inside the cage at *R*. From where it is lying, the snake can see the rodent. Trace the above diagram and identify the regions in which the snake could be lying.

Exercise 7.3

1 A coin is rolled in a straight line on a flat surface as shown below.

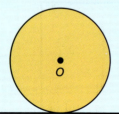

Draw the locus of the centre of the coin *O* as the coin rolls along the surface.

2 The diameter of the disc below is the same as the width and height of each of the steps shown.

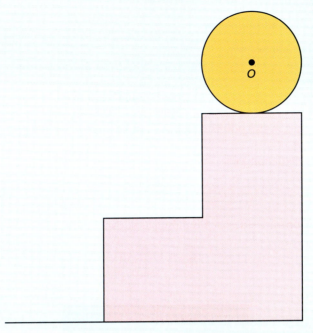

Copy the diagram and draw the locus of the centre of the disc as it rolls down the steps.

3 A stone is thrown at an angle of elevation of 45°. Sketch the locus of the stone. (**Note.** The angle of elevation is the angle made with the horizontal.)

4 X and Y are two fixed posts in the ground. The ends of a rope are tied to X and Y. A goat is attached to the rope by a ring on its collar which enables it to move freely along the rope's length.

Copy the diagram above and sketch the locus of points in which the goat is able to graze.

SUMMARY

By the end of this chapter you should know:

■ what is meant by the term **locus/loci**
■ how to construct the locus of the points which are at a given distance from a point (a circle)
■ how to construct the locus of the points which are at a given distance from a straight line
■ how to construct the locus of the points which are **equidistant** from two given points (the **perpendicular bisector**)
■ how to construct the locus of the points which are equidistant from two given intersecting straight lines (the angle bisector)
■ how to solve problems involving loci.

Exercise 7A

1 Pedro and Sara are on opposite sides of a building as shown below.

building

Their friend Raul is standing in a place such that he cannot be seen by either Pedro or Sara. Copy the above diagram and identify the locus of points in which Raul could be standing.

2 A rectangular rose bed in a park measures 8 m by 5 m as shown below.

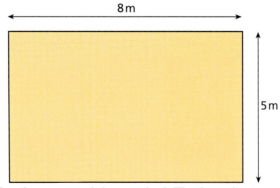

The park keeper puts a low fence around the rose bed. The fence is at a constant distance of 2 m from the rose bed.
a Draw a scale diagram of the rose bed.
b Draw the position of the fence.

3 *A* and *B* are two radio beacons 80 km apart at either side of a shipping channel. A ship sails in such a way that it is always equidistant from *A* and *B*.

A   80 km   B

Showing your method of construction clearly, draw the path of the ship.

4 A ladder 10 m long is propped up against a wall as shown. A point *P* on the ladder is 2 m from the top. Draw a scale diagram to show the locus of point *P* if the ladder were to slide down the wall.

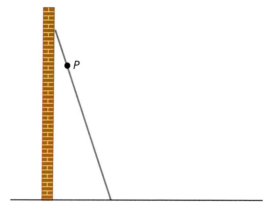

Note. Several positions of the ladder will need to be shown.

5 The equilateral triangle *PQR* is rolled along the line shown. At first, corner *Q* acts as the pivot point until *P* reaches the line, then *P* acts as the pivot point until *R* reaches the line, and so on.

Showing your method clearly, draw the locus of point *P* as the triangle makes one full rotation.

Exercise 7B

1 José, Katrina and Luis are standing at different points around a building as shown below.

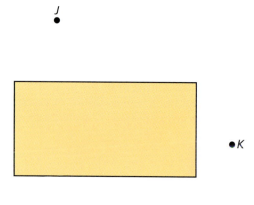

Trace the above diagram and show whether any of the three friends can see each other or not.

2 A rectangular courtyard measures 20 m by 12 m. A horse is tethered in the centre with a rope 7 m long. Another horse is tethered, by a rope 5 m long, to a rail which runs along the whole of the left-hand side of the courtyard. This rope is able to run freely along the length of the rail. Draw a scale diagram of the courtyard and draw the locus of points which can be reached by both horses.

3 The view of the diagram below is of two walls which form part of an obstacle course. A girl decides to ride her bicycle in between the two walls in such a way that she is always equidistant from them.

Copy the diagram above and, showing your construction clearly, draw the locus of her path.

4 A ball is rolling along the line shown in the diagram below.

Copy the above diagram and draw on the locus of the centre *O* of the ball as it rolls.

Exercise 7C Ma1

You will need:
• computer with Cabri II or similar software installed

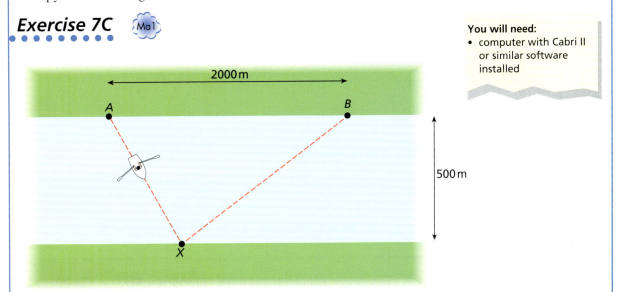

The little village of Watermead has an annual rowing race. Contestants have to row from point *A* on one side of the river, to the opposite bank and then back to point *B*. The rules state that a rower can turn at any point on the opposite bank – i.e. point *X* can be anywhere along the opposite river bank.

Using Cabri or a similarly geometry package, work out the shortest possible distance that a rower can row. Set up your screen similarly to the one below:

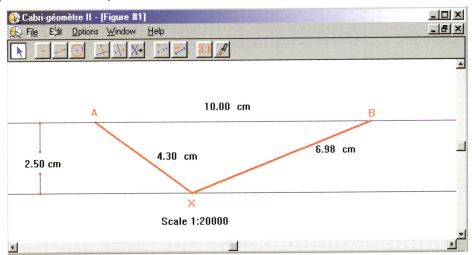

Exercise 7D

Look at the question in exercise 7C. Use a spreadsheet to model the problem and find the shortest distance $AX + XB$ correct to 2 d.p.

A simplified diagram of the information is given below:

You will need:
- computer with spreadsheet package installed

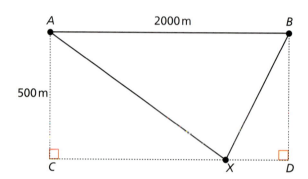

Set up your spreadsheet in a similar way to the one shown below:

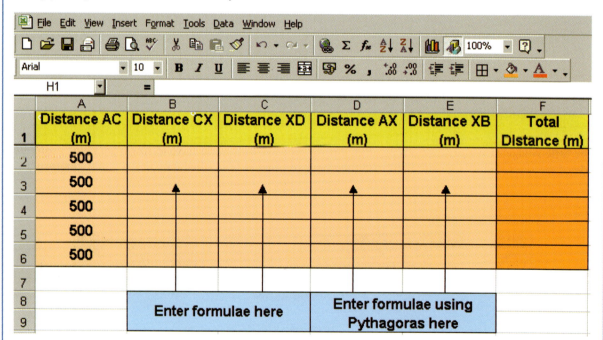

Exercise 7E

You will need:
- computer with internet access

Helium, argon, krypton, xenon and radon are known as the noble gases. Using the internet as a resource, find out which elements are known as the noble metals. How are the noble metals similar to the noble gases?

8 Mensuration

(A calculator is essential for this chapter.)

Egyptian society around 2000 BC was very advanced, particularly in its understanding and development of new mathematical ideas and concepts. One of the most important pieces of Egyptian work is called the 'Moscow Papyrus' – so called because it was taken to Moscow in the middle of the nineteenth century. It was written in about 1850 BC and is important because it contains 25 mathematical problems. One of the key problems and its solution is finding the volume of a truncated pyramid.

Although the solution was not written in the way we write it today, it was mathematically correct and translates into the formula:

$$V = \frac{(a^2 + ab + b^2)h}{3}$$

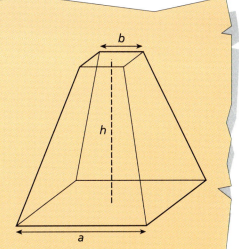

Perimeter, area and volume

(*Revision*)
You will already have learned about the area and perimeter of rectangles and squares. You will also have found the area of a triangle given the length of the base and the perpendicular height.

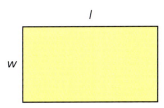

perimeter $= 2l + 2w$
area $= lw$

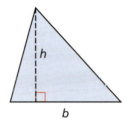

area $= \frac{1}{2}bh$

You should also be aware that the perimeter of a circle is called its **circumference** and that the circumference and area of a circle can be calculated if its radius is given.

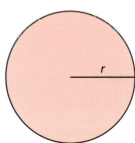

circumference $= 2\pi r$
area $= \pi r^2$

You may know that a **prism** is defined as a *three-dimensional solid object with a constant **cross-section***. Examples are shown below:

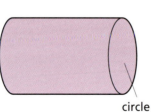
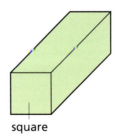
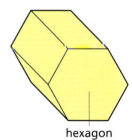

circle   square   hexagon

Exercise 8.1

(*Revision*)
1 Calculate the area and perimeter of each of the rectangles described below.

	length	width	area	perimeter
a	6 cm	4 cm		
b	3.8 m	10 m		
c	3.2 cm	4.7 cm		
d	85 cm	1.2 m		
e	3.3 m	75 cm		

2 Use the formula for the area of a rectangle to work out the missing values in the table below.

	length	width	area
a	8.5 cm	7.2 cm	
b	25 cm		250 cm²
c		25 cm	400 cm²
d		7.5 cm	187.5 cm²

3 Calculate the area of each of the triangles below.

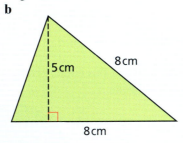
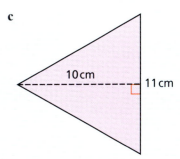

4 Calculate the area of each of the shapes below.

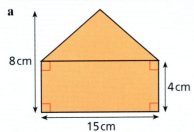
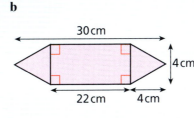
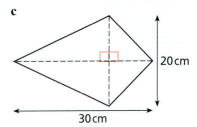

Circumference and area of a circle

Remember:
$\pi = 3.142$ to 3 d.p.

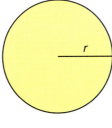
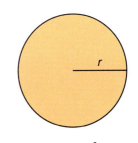

circumference $= 2\pi r$

area $= \pi r^2$

Examples   Calculate the circumference of this circle giving your answer to 2 d.p.

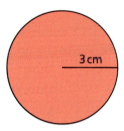

3 cm

circumference $= 2\pi r$
$= 2\pi \times 3$
$= 18.85$

The circumference is 18.85 cm.

Calculate the radius of this circle, giving your answer to 2 d.p., if its circumference is 12 cm.

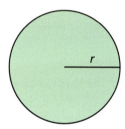

r

$C = 2\pi r$

$r = \dfrac{C}{2\pi}$

$r = \dfrac{12}{2\pi}$

$r = 1.91$

The radius is 1.91 cm.

Calculate the area of this circle giving your answer to 2 d.p.

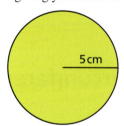

5 cm

area $= \pi r^2$
$= \pi \times 5^2$
$= 78.54$

The area is 78.54 cm^2.

Calculate the radius of this circle, giving your answer to 2 d.p., if its area is 34 cm².

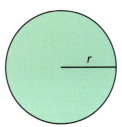

$$A = \pi r^2 \qquad \text{Therefore} \quad \frac{A}{\pi} = r^2$$

$$r = \sqrt{\frac{A}{\pi}}$$

$$r = \sqrt{\frac{34}{\pi}}$$

$$r = 3.29$$

The radius is 3.29 cm.

Exercise 8.2

1 Calculate the circumference of each circle, giving your answer to 2 d.p.

a **b** **c** **d**

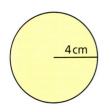
4 cm

3.5 cm

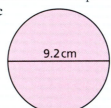
9.2 cm

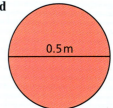
0.5 m

2 Calculate the area of each of the circles in question 1. Give your answers to 2 d.p.
3 Calculate the radius of a circle when the circumference is:
    a 15 cm   **b** π cm   **c** 4 cm   **d** 8 mm
4 Calculate the diameter of a circle when the area is:
    a 16 cm²   **b** 9π cm²   **c** 8.2 m²   **d** 14.6 mm²

Exercise 8.3

1 The wheel of a car has an outer radius of 25 cm. Calculate:
    a how far the car travels for one complete turn of the wheel
    b how many times the wheel will rotate in a 1 km journey.

2 If the wheel of a bicycle has a diameter of 60 cm, calculate how far a cyclist will have travelled after the wheel has rotated 100 times.

3 A circular ring (an annulus) has a cross-section as shown. If the outer radius is 22 mm and the inner radius 20 mm, calculate the cross-sectional area of the ring.

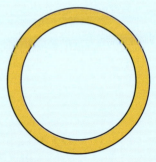

4

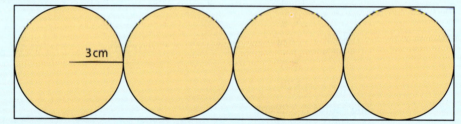

Four circles are drawn in a line and enclosed by a rectangle as shown. If the radius of each circle is 3 cm, calculate:

a the area of the rectangle

b the area of each circle

c the unshaded area of the rectangle.

5 A garden is made up of a rectangular patch of grass and two semi-circular vegetable patches. If the length and width of the rectangular patch are 16 m and 8 m respectively, calculate:

a the perimeter of the garden

b the total area of the garden.

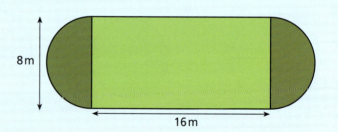

Area of parallelograms and trapeziums

A parallelogram can be rearranged to form a rectangle in the following way:

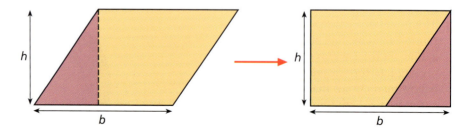

Therefore:

$$\text{area of parallelogram} = \text{base length} \times \text{perpendicular height}$$

A trapezium can be visualised as being split into two triangles as follows:

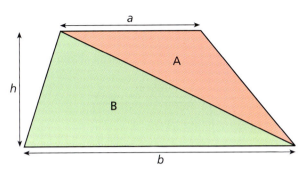

area of triangle A $= \frac{1}{2} \times a \times h$
area of triangle B $= \frac{1}{2} \times b \times h$

The area of the trapezium = area of triangle A + area of triangle B

$$= \frac{1}{2}ah + \frac{1}{2}bh$$
$$= \frac{1}{2}h(a + b)$$

i.e. $\frac{1}{2} \times$ the sum of the parallel sides $\times$ the distance between them.

Examples Calculate the area of the parallelogram.

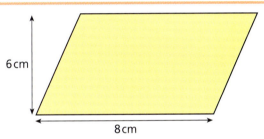

area = base length $\times$ perpendicular height
      $= 8 \times 6$
      $= 48$

The area is $48 \, cm^2$.

Calculate the area of the shaded region in the following shape:

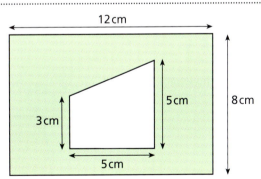

area of rectangle $\; = 12 \times 8$
$\qquad\qquad\qquad = 96$
area of trapezium $= \frac{1}{2} \times 5(3 + 5)$
$\qquad\qquad\qquad = 2.5 \times 8$
$\qquad\qquad\qquad = 20$
shaded area $\qquad = (96 - 20)$
$\qquad\qquad\qquad = 76$

The area of the shaded region is $76 \, cm^2$.

Exercise 8.4

Find the area of each of the following shapes.

1

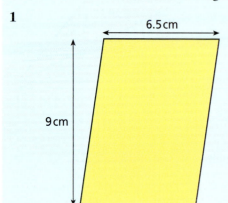

6.5 cm

9 cm

2

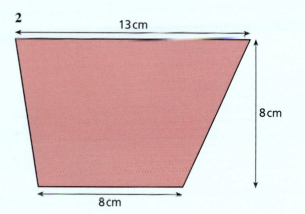

13 cm

8 cm

8 cm

3

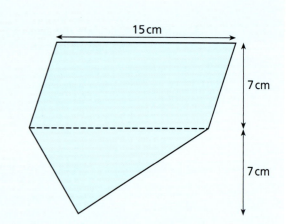

11 cm

7.2 cm

6.4 cm

4

15 cm

7 cm

7 cm

Exercise 8.5

1 Calculate the length marked *a*.

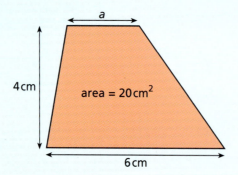

a

4 cm

area = 20 cm²

6 cm

2 If the areas of the trapezium and parallelogram are equal, calculate the length marked x.

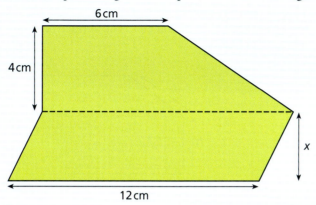

3 The end view of a house is as shown in the diagram. If the door has a width and height of 0.75 m and 2 m respectively and the window has a diameter of 0.8 m, calculate the area of brickwork.

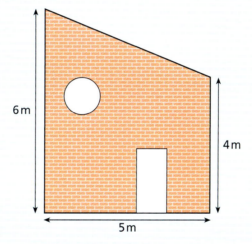

4 A garden in the shape of a trapezium is split into three parts: two flower beds in the shape of a triangle and a parallelogram, and a section of grass in the shape of a trapezium. The area of the grass is two and a half times the area of the flower beds added together. Calculate:
a the area of each flower bed
b the area of grass
c the length marked x.

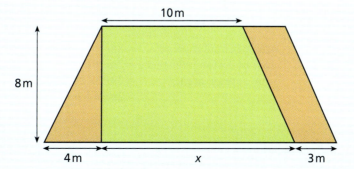

Surface area of cuboids and cylinders

To calculate the surface area of a cuboid, we look at its individual faces. These are either squares or rectangles. The surface area of a cuboid is the sum of the areas of its faces.

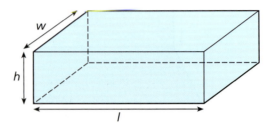

area of top and bottom $= 2wl$
area of front and back $= 2lh$
area of both sides $\quad = 2wh$
total surface area $\qquad = 2wl + 2lh + 2wh$
$\qquad\qquad\qquad\quad\; = 2(wl + lh + wh)$

For the surface area of a cylinder, it is best to visualise the net of the solid: it is made up of one rectangular piece and two circular pieces.

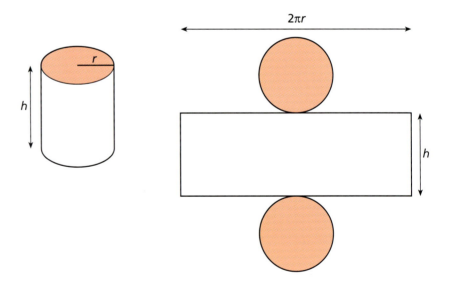

Remember:
The width of the rectangle is the same as the circumference of the circle.

area of circular pieces $\quad = 2\pi r^2$
area of rectangular piece $= 2\pi r \times h$
total surface area $\qquad\quad = 2\pi r^2 + 2\pi rh$
$\qquad\qquad\qquad\qquad\; = 2\pi r(r + h)$

Examples   Calculate the surface area of the cuboid shown below.

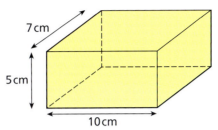

area of top and bottom $= 2 \times 7 \times 10\,\text{cm}^2 = 140\,\text{cm}^2$
area of front and back $= 2 \times 5 \times 10\,\text{cm}^2 = 100\,\text{cm}^2$
area of both sides $= 2 \times 5 \times 7\,\text{cm}^2 = 70\,\text{cm}^2$

Therefore the total surface area $= (140 + 100 + 70)\,\text{cm}^2 = 310\,\text{cm}^2$.

If the height of a cylinder is 7 cm and the radius of its circular top is 3 cm, calculate the surface area.

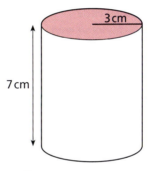

total surface area $= 2\pi r^2 + 2\pi rh$
$= 2\pi r(r + h)$
$= 2\pi \times 3 \times (3 + 7)$
$= 6\pi \times 10$
$= 60\pi$
$= 188.50$ (2 d.p.)

Remember:
Factorise whenever you can!

The total surface area is $188.50\,\text{cm}^2$.

Exercise 8.6

1 Calculate the surface area of each of the following cuboids, given the length, width and height.
  a $l = 12\,\text{cm}$, $w = 10\,\text{cm}$, $h = 5\,\text{cm}$ **b** $l = 4\,\text{cm}$, $w = 6\,\text{cm}$, $h = 8\,\text{cm}$
  c $l = 4.2\,\text{cm}$, $w = 7.1\,\text{cm}$, $h = 3.9\,\text{cm}$ **d** $l = 5.2\,\text{cm}$, $w = 2.1\,\text{cm}$, $h = 0.8\,\text{cm}$
2 Calculate the height of each of the following cuboids, giving the length, width and surface area.
  a $l = 5\,\text{cm}$, $w = 6\,\text{cm}$,   surface area $= 104\,\text{cm}^2$ **b** $l = 2\,\text{cm}$, $w = 8\,\text{cm}$,   surface area $= 112\,\text{cm}^2$
  c $l = 3.5\,\text{cm}$, $w = 4\,\text{cm}$,   surface area $= 118\,\text{cm}^2$ **d** $l = 4.2\,\text{cm}$, $w = 10\,\text{cm}$,   surface area $= 226\,\text{cm}^2$
3 Calculate the surface area of each of the following cylinders, given the radius and height.
  a $r = 2\,\text{cm}$, $h = 6\,\text{cm}$ **b** $r = 4\,\text{cm}$, $h = 7\,\text{cm}$
  c $r = 3.5\,\text{cm}$, $h = 9.2\,\text{cm}$ **d** $r = 0.8\,\text{cm}$, $h = 4.3\,\text{cm}$
4 Calculate the height of each of the following cylinders. Give your answers to 1 d.p.
  a $r = 2.0\,\text{cm}$,   surface area $= 40\,\text{cm}^2$ **b** $r = 3.5\,\text{cm}$,   surface area $= 88\,\text{cm}^2$
  c $r = 5.5\,\text{cm}$,   surface area $= 250\,\text{cm}^2$ **d** $r = 3.0\,\text{cm}$,   surface area $= 189\,\text{cm}^2$

Exercise 8.7

1

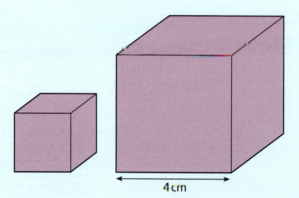

Two cubes are placed next to each other. The length of each of the edges of the larger cube is 4 cm. If the ratio of their surface areas is 1:4, calculate:
a the surface area of the small cube
b the length of an edge of the small cube.

2 A cube and a cylinder have the same surface area. If the cube has an edge length of 6 cm and the cylinder a radius of 2 cm, calculate:
a the surface area of the cube
b the height of the cylinder.

3 Two cylinders have the same surface area.

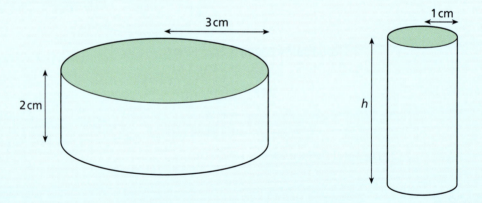

The shorter of the two has a radius of 3 cm and a height of 2 cm, while the taller cylinder has a radius of 1 cm. Calculate:
a the surface area of one of the cylinders
b the height of the taller cylinder.

4 Two cuboids have the same surface area. The dimensions of one of them are length 3 cm, width 4 cm and height 2 cm.
Calculate the height of the other cuboid if its length is 1 cm and width is 4 cm.

Volume of prisms

A prism is any three-dimensional object which has a constant cross-sectional area. Below are a few examples of some of the more common types of prism:

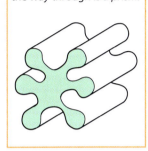

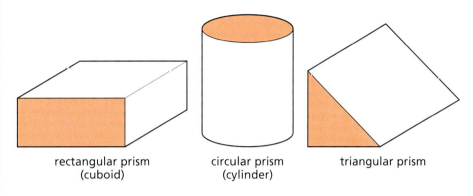

rectangular prism (cuboid)   circular prism (cylinder)   triangular prism

When each of the objects is cut parallel to the shaded faces, the cross-section is constant and therefore the object is classified as a prism.

volume of a prism = area of cross-section × length

Examples   Calculate the volume of this cylinder.

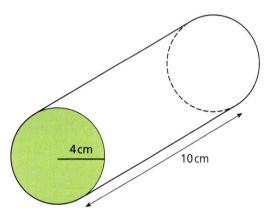

4 cm

10 cm

$$\text{volume} = \text{cross-sectional area} \times \text{length}$$
$$= \pi \times 4^2 \times 10$$
$$= 502.7 \ (1 \, \text{d.p.})$$

The volume of the cylinder is $502.7 \, \text{cm}^3$.

Calculate the volume of this L-shaped prism.

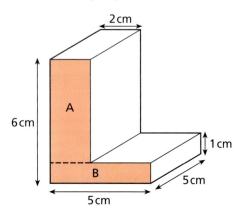

The cross-sectional area can be split into two rectangles:

$$\text{area of rectangle A} = (5 \times 2)\,\text{cm}^2$$
$$= 10\,\text{cm}^2$$
$$\text{area of rectangle B} = (5 \times 1)\,\text{cm}^2$$
$$= 5\,\text{cm}^2$$
$$\text{total cross-sectional area} = (10 + 5)\,\text{cm}^2 = 15\,\text{cm}^2$$
$$\text{volume of prism} = (15 \times 5)\,\text{cm}^3$$
$$= 75\,\text{cm}^3$$

The volume of the prism is $75\,\text{cm}^3$.

Exercise 8.8

1 Calculate the volume of each of the following cuboids, where w, l and h represent the width, length and height respectively.

  a $w = 2\,\text{cm}$, $l = 3\,\text{cm}$, $h = 4\,\text{cm}$
  b $w = 6\,\text{cm}$, $l = 1\,\text{cm}$, $h = 3\,\text{cm}$
  c $w = 6\,\text{cm}$, $l = 23\,\text{mm}$, $h = 2\,\text{cm}$
  d $w = 42\,\text{mm}$, $l = 3\,\text{cm}$, $h = 0.007\,\text{m}$

2 Calculate the volume of each of the following cylinders, where r represents the radius of the circular face and h the height of the cylinder respectively.

  a $r = 4\,\text{cm}$, $h = 9\,\text{cm}$
  b $r = 3.5\,\text{cm}$, $h = 7.2\,\text{cm}$
  c $r = 25\,\text{mm}$, $h = 10\,\text{cm}$
  d $r = 0.3\,\text{cm}$, $h = 17\,\text{mm}$

3 Calculate the volume of each of the following triangular prisms, where b represents the base length of the triangular face, h its perpendicular height and l the length of the prism.

  a $b = 6\,\text{cm}$, $h = 3\,\text{cm}$, $l = 12\,\text{cm}$
  b $b = 4\,\text{cm}$, $h = 7\,\text{cm}$, $l = 10\,\text{cm}$
  c $b = 5\,\text{cm}$, $h = 24\,\text{mm}$, $l = 7\,\text{cm}$
  d $b = 62\,\text{mm}$, $h = 2\,\text{cm}$, $l = 0.01\,\text{m}$

4 Calculate the volume of each of the following prisms. All dimensions are given in centimetres.

a

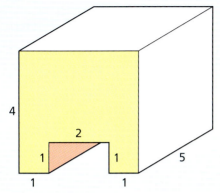

b

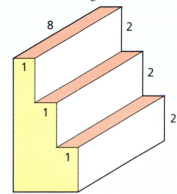

c

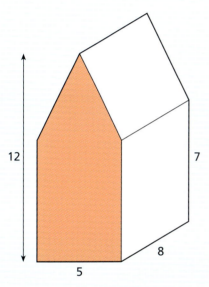

d

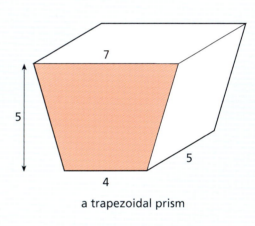

a trapezoidal prism

Exercise 8.9

1 Below is a plan view of a cylinder inside a box the shape of a cube. If the radius of the cylinder is 8 cm, calculate:
   a the height of the cube
   b the volume of the cube
   c the volume of the cylinder
   d the percentage volume of the cube not occupied by the cylinder.

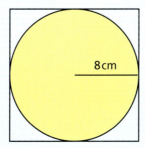

8 cm

2 A chocolate bar is made in the shape of a triangular prism. The triangular face of the prism is equilateral and has an edge length of 4 cm and a perpendicular height of 3.5 cm. The manufacturer also sells these in special packs of six bars arranged as a hexagonal prism. If the prisms are 20 cm long, calculate:
a the cross-sectional area of the pack
b the volume of the pack.

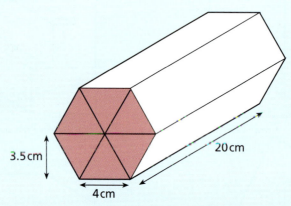

3 A cuboid and a cylinder have the same volume. The radius and height of the cylinder are 2.5 cm and 8 cm respectively. If the length and width of the cuboid are both 5 cm, calculate its height to 1 d.p.
4 A section of steel pipe is shown in the diagram below. The inner radius is 35 cm and the outer radius is 36 cm. Calculate the volume of steel used in making the pipe if it has a total length of 130 m.

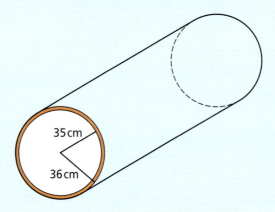

Dimensions

Consider the formula for the perimeter of a rectangle.

$$P = l + w + l + w$$
$$= 2l + 2w$$

Since l and w are lengths, it is clear that the perimeter is a length (adding lengths gives a bigger length), which has *one* **dimension**.

The area of a rectangle is given by the formula $A = lw$. Area has therefore *two dimensions* as it has two lengths multiplied together, giving a measurement in **units squared** (e.g. cm²).

A cuboid has *three dimensions* (length, width and height multiplied together), giving a measurement in **units cubed** (e.g. cm³) – the volume.

> **Remember:**
> *A line, one dimension*
> *mm, cm, m*
> *A surface, two dimensions*
> *mm², cm², m²*
> *A solid, three dimensions*
> *mm³, cm³, m³*

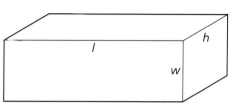

It is therefore possible to deduce what a quantity is measuring by looking at its units.

Area is measured in m², cm², mm², etc. and has *two dimensions* (for a rectangle these would be length and width).

Volume is measured in m³, cm³, mm³, etc. and has *three dimensions* (for a cuboid these would be length, width and height).

> A formula with a mixture of things doesn't mean anything. For example
>
> $$\underset{\downarrow}{\pi r^2} + \underset{\downarrow}{\tfrac{4}{3}\pi r^3}$$
> $$\text{area} + \text{volume}$$
>
> is meaningless.

Note. It is important to realise that π represents a number 3.141…, *not* a dimension. If this is understood then it is clear that:

$C = 2\pi r$ represents a length (in this case the circumference of a circle)
$A = \pi r^2$ represents an area (the area of a circle)
$V = \tfrac{4}{3}\pi r^3$ represents a volume (the volume of a sphere)

Exercise 8.10

By considering the formulae below, state whether they could be formulae for length, area or volume. (**Note.** Each of the lower case letters represents a length.)

1 $L = a + b + c$
2 $A = bc$
3 $V = pqr$
4 $M = s + t - v$
5 $W = pq$
6 $S = abc$
7 $V = 5t$
8 $T = \pi x$
9 $M = \pi m^2$
10 $W = \pi a^2 b$

> **Remember:**
> *5, π, $\tfrac{4}{3}$ etc. don't affect the dimensions of a formula.*

Converting units

Converting from one metric unit to another usually involves multiplying or dividing by a power of 10, for example

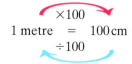

$$1 \text{ metre} = 100 \text{ cm}$$

> **Remember:**
> $10^2 = 100$

to convert metres to cm multiply by 100
to convert cm to metres divide by 100

Dealing with conversions involving areas and volumes tends to cause more problems but it needn't.

In the diagram below, the square has a side length of 1 m (i.e. 100 cm).

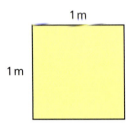

1 m

1 m

$$\text{area} = 1\,\text{m} \times 1\,\text{m} = 1\,\text{m}^2$$
$$\text{area} = 100\,\text{cm} \times 100\,\text{cm} = 10\,000\,\text{cm}^2$$
$$\text{therefore } 1\,\text{m}^2 = 10\,000\,\text{cm}^2$$

Remember:
$10^4 = 10\,000$

$$\overset{\times 10\,000}{\underset{\div 10\,000}{1\,\text{m}^2 \quad = \quad 10\,000\,\text{cm}^2}}$$

i.e. to convert m^2 to cm^2 multiply by 10 000
to convert cm^2 to m^2 divide by 10 000

The cube below has a side length of 1 m.

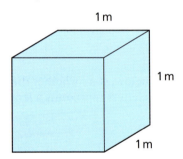

1 m

1 m

1 m

$$\text{volume} = 1\,\text{m} \times 1\,\text{m} \times 1\,\text{m} = 1\,\text{m}^3$$
$$\text{volume} = 100\,\text{cm} \times 100\,\text{cm} \times 100\,\text{cm}$$
$$= 1\,000\,000\,\text{cm}^3$$
$$\text{therefore } 1\,\text{m}^3 = 1\,000\,000\,\text{cm}^3$$

Remember:
$10^6 = 1\,000\,000$

$$\overset{\times 1\,000\,000}{\underset{\div 1\,000\,000}{1\,\text{m}^3 \quad = \quad 1\,000\,000\,\text{cm}^3}}$$

i.e. to convert m^3 to cm^3 multiply by 1 000 000
to convert cm^3 to m^3 divide by 1 000 000

Exercise 8.11

Convert the following, by copying and filling in the missing quantities.

1 $2.7\,\text{m} = \ldots\text{cm}$ **2** $8\,\text{m}^2 = \ldots\text{cm}^2$ **3** $3.1\,\text{m}^3 = \ldots\text{cm}^3$ **4** $84\,000\,\text{cm}^2 = \ldots\text{m}^2$

5 $7\,800\,000\,\text{cm}^3 = \ldots\text{m}^3$ **6** $5500\,\text{cm}^2 = \ldots\text{m}^2$ **7** $80\,000\,\text{mm}^2 = \ldots\text{cm}^2$ **8** $9.2\,\text{m}^2 = \ldots\text{mm}^2$

9 $3.5\,\text{cm}^3 = \ldots\text{mm}^3$ **10** $2\,\text{m}^2 = \ldots\text{cm}^2$

SUMMARY

By the time you have completed this chapter you should know:

■ how to calculate the area and perimeter of simple shapes by using the appropriate formulae
■ how to find the **circumference** and area of a circle using the appropriate formulae

$$C = 2\pi r \qquad A = \pi r^2$$

■ how to solve problems involving the surface area and volume of **prisms**, including cuboids and cylinders, and shapes made from cubes and cuboids
■ the difference between formulae for perimeter, area and volume by considering **dimensions**
■ how to convert between measures including cm and m, cm^2 and m^2 and cm^3 and m^3.

Exercise 8A

1 A swimming pool is 50 m long by 20 m wide.
  a Find the surface area and perimeter of the pool.
  b Square tiles of side length 25 cm are placed around the edge and corners of the pool. How many tiles will be needed to fit around it?
2 A floor measures 8 m by 6 m and is to be covered in square tiles of side 50 cm. How many tiles are needed?
3 A carpet is 2 m 40 cm by 3 m 80 cm. Calculate its area and perimeter.
4 A setsquare has a base length of 40 cm and a height of 25 cm. Calculate its area.
5 a Calculate the area of the shape below:

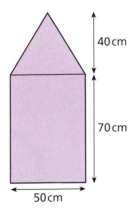

40 cm

70 cm

50 cm

  b Calculate the area of the above shape if each of the lengths were doubled.

Exercise 8B

1 Calculate the area and circumference of each of the following circles. Give your answers correct to 1 d.p.
  a

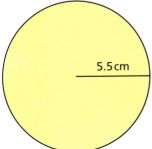

5.5 cm

  b

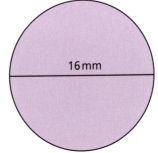

16 mm

2 A semi-circular shape is cut out of the side of a rectangle as shown. Calculate the shaded area correct to 1 d.p.

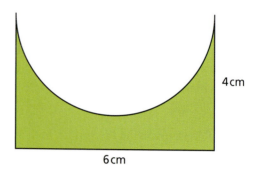

4 cm

6 cm

3 For the shape shown in the diagram, calculate the area of:
   a the semi-circle
   b the trapezium
   c the whole shape.

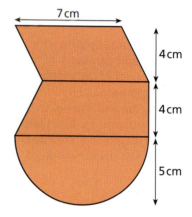

7 cm

4 cm

4 cm

5 cm

4 A cylindrical tube has an inner diameter (a bore) of 6 cm, an outer diameter of 7 cm and a length of 15 cm.
Calculate:
   a the surface area of the ring-shaped end
   b the inside surface area of the tube
   c the total surface area of the tube.

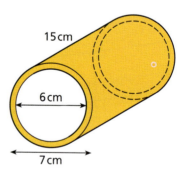

15 cm

6 cm

7 cm

5 Calculate the volume and surface area of the cylinder shown.

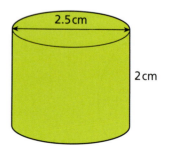

2.5 cm

2 cm

Exercise 8C

You will need:
• computer with spreadsheet package installed

A supermarket chain is planning to bring out a new tin of baked beans with a volume of $1000\,cm^3$.

a Write the formula for the volume V of the cylinder in terms of r and h.
b Rearrange the formula for the volume of the cylinder to make h the subject.
c Write a formula for the total surface area A of the tin in terms of r and V.
d Using a spreadsheet, work out the minimum possible surface area of the tin.
e What radius is needed to produce the tin with the minimum surface area? Give your answer correct to 1 d.p.
f Why might the supermarket be interested in knowing what is the minimum possible surface area?

Exercise 8D

You will need:
• computer with Cabri II or similar software installed

Use a geometry package such as Cabri to investigate the relationship between the circumference and diameter of circles.

• Draw several circles on the screen. For each one, get the computer to calculate the circumference and diameter.
• Tabulate the results and search for any relationship between the two variables.

Your screen may look similar to the one shown below:

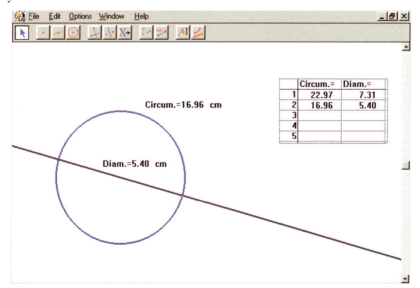

Exercise 8E

You will need:
• computer with internet access

Samuel Plimsoll (1824–1898) became aware that many merchant ships sank because they were overloaded. This was either deliberate, in an attempt to make profits (or claim insurance), or through ignorance of the fact that some seas were more buoyant than others.

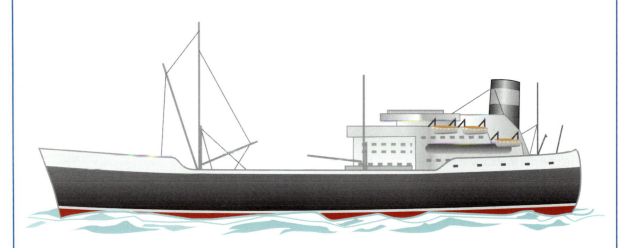

Using the internet as a resource, find out and illustrate what Plimsoll devised to show the maximum load that a ship is allowed to carry.

9 Pythagoras and trigonometry

Trigonometry and the trigonometric ratios developed from the ancient study of the stars. The study of right-angled triangles probably originated with the Egyptians and the Babylonians, who used these triangles extensively in construction and engineering. The trigonometric ratios, which are introduced in this chapter, were set out by Hipparchus of Rhodes about 150 BC.

Trigonometry was used extensively in navigation at sea particularly in the sailing ships of the eighteenth and nineteenth centuries, when it formed a major part of the examination to become a lieutenant in the Royal Navy.

Hipparchus discovered the trigonometric ratios

Lord Nelson would have used trigonometry in navigation

> Trigonometry is still used in navigation at sea and now also in the air, and on land by the Global Positioning System (GPS).

Bearings

(*Revision*)

In the days when sailing ships travelled the oceans of the world, compass directions like the ones in this diagram were used.

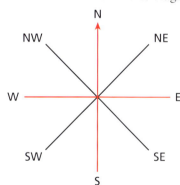

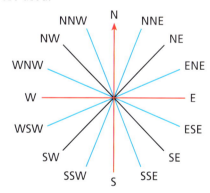

As the need for more accurate directions arose, extra points were added to the traditional eight-point compass. Midway between north and north east was north north east and so on. This gave 16 points which was later extended to 32 points and still further to 64 points.

As the speed of travel increased and greater accuracy was needed, a new system was required. The new system was the **three-figure bearing** system. North was given the bearing 000°. One full rotation in a clockwise direction was equivalent to 360°.

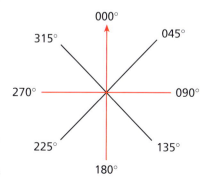

> **Remember:**
> *Bearings are always measured clockwise from north.*

Exercise 9.1

You will need:
• ruler
• protractor

(*Revision*)

Draw diagrams to show the following bearings and journeys. Use a scale of 1 cm : 1 km. North can be taken to be a vertical line up the page.

Remember:
1 cm : 1 km
is
1 : 100 000

1 Start at a point A. Travel a distance of 7 km on a bearing of 135° to point B. From B, travel 12 km on a bearing of 250° to point C. Measure the distance and bearing of A from C.

2 Start at a point P. Travel a distance of 6.5 km on a bearing of 225° to point Q. From Q travel 7.8 km on a bearing of 105° to point R. From R, travel 8.5 km on a bearing of 090° to point S. What is the distance and bearing of P from S?

3 Start at a point P. Travel a distance of 11.2 km on a bearing of 270° to point Q. From Q, travel 5.8 km on a bearing of 170° to point O. What is the bearing and distance of P from O?

Back bearings

Examples In the diagram below, the bearing of B from A is 135°. The bearing of A from B is known as the **back bearing**. Calculate the bearing of A from B.

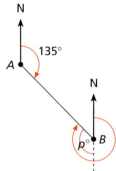

Since the two north lines are parallel, $p = 135°$ (alternate angles).
Therefore the back bearing is $180° + 135° = 315°$.

In the diagram below, the bearing of B from A is 245°. Calculate the bearing of A from B.

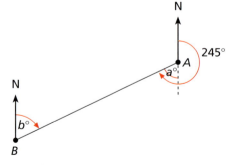

$$a = 245° - 180° = 65°$$

Since the two north lines are parallel, $a = b$ (alternate angles).
Therefore the bearing of A from B is 065°.

Exercise 9.2

You will need:
• ruler
• protractor

1 Investigate the relationship between a bearing and its back bearing.
  a Draw several pairs of points and label each pair *A* and *B* as in the worked examples above.
  b For each pair of points, measure the bearing of *B* from *A* and the back bearing of *A* from *B*.
  c Can you see any rule relating a bearing and its back bearing?
2 Given the following bearings of point *Y* from point *X*, draw diagrams and use them to *calculate* the bearings of *X* from *Y*.
  a 130°   **b** 145°   **c** 220°   **d** 163°
3 Given the following bearings of point *D* from point *C*, draw diagrams and use them to *calculate* the bearings of *C* from *D*.
  a 300°   **b** 320°   **c** 290°   **d** 351°

Trigonometry

There are three basic trigonometric ratios: **sine**, **cosine** and **tangent**. Each of these expresses an angle of a right-angled triangle as a ratio of the length of two of the sides of the triangle.

The sides of the triangle have names, two of which are dependent on their position in relation to a specific angle.

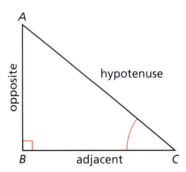

The longest side (always opposite the right angle) is called the **hypotenuse**. The side opposite the chosen angle (angle *C* in this diagram) is called the **opposite**. The side next to the angle is called the **adjacent**.

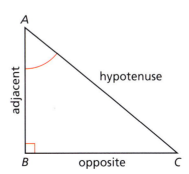

Note that when the chosen angle is at *A*, the sides labelled **opposite** and **adjacent** change.

Tangent

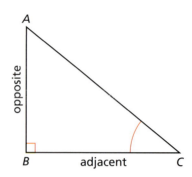

$$\tan C = \frac{\text{length of opposite side}}{\text{length of adjacent side}}$$

Examples Calculate the size of angle A in each of the following triangles.

a

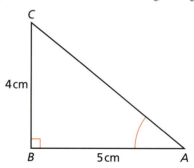

Remember:
When doing calculations with a calculator make sure it is in degree mode.

- - - - - - - - - - - -
$\tan^{-1} x$ is **not** the same

as $\dfrac{1}{\tan x}$

$A = \tan^{-1}\frac{4}{5}$ means A is the angle whose tan is $\frac{4}{5}$ or 0.8.
- - - - - - - - - - - -

a $\tan A = \dfrac{\text{opposite}}{\text{adjacent}} = \frac{4}{5}$

$A = \tan^{-1}\frac{4}{5}$
$A = 38.7°$ (1 d.p.)

Be careful: you will probably need

shift tan value =

but check your calculator to make sure:

shift tan (4 ÷ 5) =

or

shift tan . 8 =

b

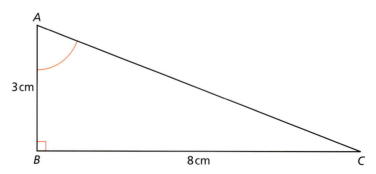

b $\tan A = \frac{8}{3}$
$A = \tan^{-1}\frac{8}{3}$
$A = 69.4°$ (1 d.p.)

Calculate the length of the opposite side:

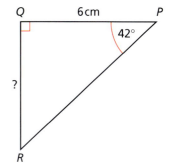

$$\tan 42° = \frac{QR}{6}$$

$$6 \times \tan 42° = QR$$
$$QR = 5.4\,\text{cm} \quad (1\,\text{d.p.})$$

Calculate the length of the adjacent side:

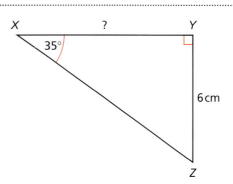

$$\tan 35° = \frac{6}{XY}$$

$$XY \times \tan 35° = 6$$

$$XY = \frac{6}{\tan 35°}$$

$$XY = 8.6\,\text{cm} \quad (1\,\text{d.p.})$$

Exercise 9.3

1 Calculate the length of the side marked ? in each of the following. Give your answers correct to 1 d.p.

a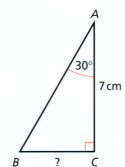

b

c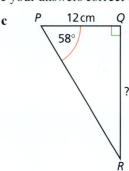

d

e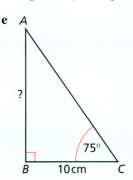

f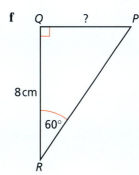

2 Calculate the length of the side marked ? in each of the following. Give your answers correct to 1 d.p.

a
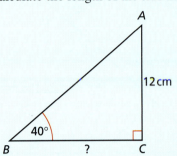

b
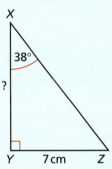

c
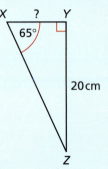

d
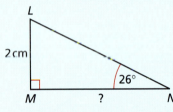

e
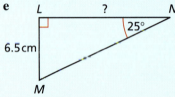

f
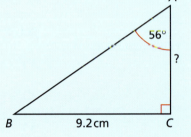

3 Calculate the size of the angle marked ? in each of the following. Give your answers correct to 1 d.p.

a
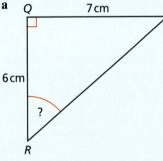

b
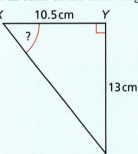

c
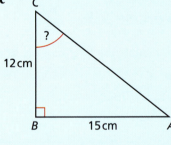

d
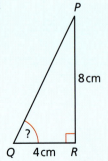

e
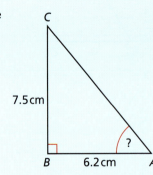

f
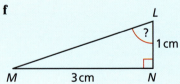

Sine

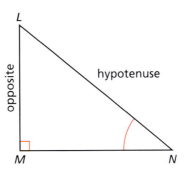

$$\sin N = \frac{\text{length of opposite side}}{\text{length of hypotenuse}}$$

Examples Calculate the size of angle X:

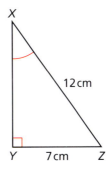

$$\sin X = \frac{\text{opposite}}{\text{hypotenuse}} = \frac{7}{12}$$

> $X = \sin^{-1}\frac{7}{12}$ means X is the angle whose sin is $\frac{7}{12}$.

$X = \sin^{-1}\frac{7}{12}$

$X = 35.7°$ (1 d.p.)

Calculate the length of the hypotenuse:

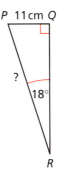

$$\sin 18° = \frac{11}{PR}$$

$$PR \times \sin 18° = 11$$

$$PR = \frac{11}{\sin 18°}$$

$$PR = 35.6 \,\text{cm} \quad (1 \,\text{d.p.})$$

Exercise 9.4

Calculate the length of the side marked ? in each of the following. Give your answers to 1 d.p.

1 a
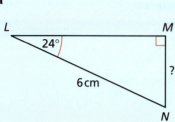

b
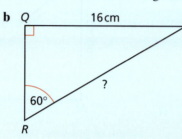

c
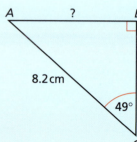

d
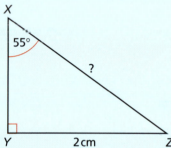

e

f
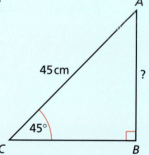

2 Calculate the size of the angle marked ? in each of the following. Give your answers to 1 d.p.

a
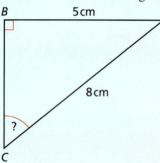

b
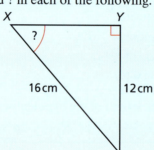

c

d
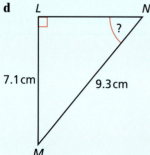

e
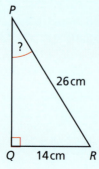

f

Cosine

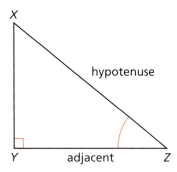

$$\cos Z = \frac{\text{length of adjacent side}}{\text{length of hypotenuse}}$$

Examples Calculate the length of XY.

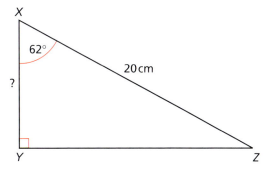

$$\cos 62° = \frac{\text{adjacent}}{\text{hypotenuse}} = \frac{XY}{20}$$

$XY = 20 \times \cos 62°$
$XY = 9.4\,\text{cm}$ (1 d.p.)

Calculate the size of angle B:

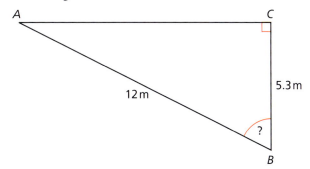

$$\cos B = \frac{5.3}{12}$$

$$B = \cos^{-1}\frac{5.3}{12}$$

$B = 63.8°$ (1 d.p.)

> $B = \cos^{-1}\dfrac{5.3}{12}$ means B is the angle whose cos is $\dfrac{5.3}{12}$.

Exercise 9.5

Calculate either the side or angle marked ? in each of the following. Give your answers to 1 d.p.

1 a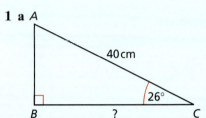

b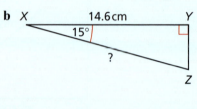

c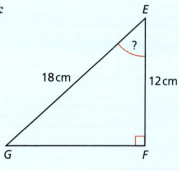

d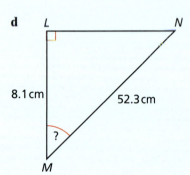

e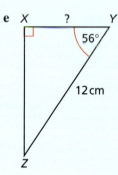

f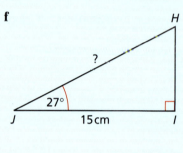

g

h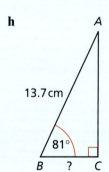

Pythagoras' theorem

Practical task

- On a piece of paper, draw a right-angled triangle.
- Off each of the sides construct a square, as shown in the diagram below.
- Split one of the smaller squares as shown, ensuring the cuts run parallel to the sides of the largest square.

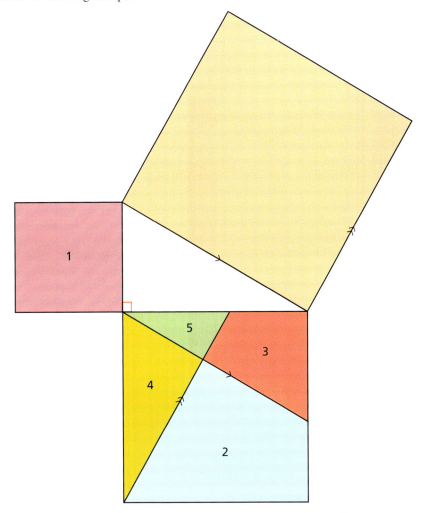

- By cutting out the shapes numbered 1, 2, 3, 4 and 5, is it possible to arrange them without any gaps on top of the largest square?
- What conclusions can you make about the areas of the three squares?

Pythagoras' theorem

● *Pythagoras' theorem states the relationship between the lengths of the three sides of a right-angled triangle:*

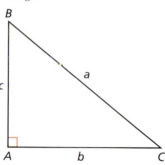

$$a^2 = b^2 + c^2$$

Note. Always start with 'the square on the hypotenuse' =.

Examples   Calculate the length of the side marked a.

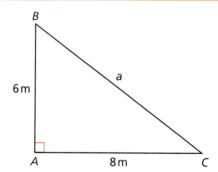

Using Pythagoras:

$a^2 = b^2 + c^2$
$a^2 = 8^2 + 6^2$
$a^2 = 64 + 36 = 100$
$a = \sqrt{100}$
$a = 10\,\text{m}$

Calculate the length of the side marked b.

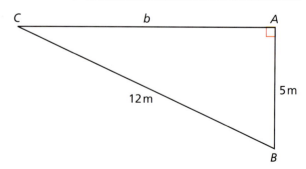

Using Pythagoras:

$a^2 = b^2 + c^2$
$a^2 - c^2 = b^2$
$b^2 = 144 - 25 = 119$
$b = \sqrt{119}$
$b = 10.9\,\text{m}$ (1 d.p.)

Coordinates of the midpoint of a line segment

In the diagram P is the midpoint of the line segment AB.

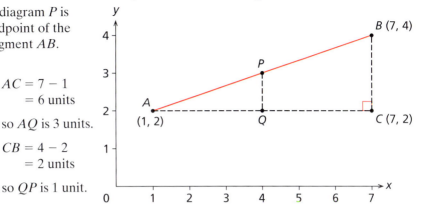

$AC = 7 - 1$
$\quad = 6$ units
so AQ is 3 units.

$CB = 4 - 2$
$\quad = 2$ units
so QP is 1 unit.

Therefore P has x-coordinate $1 + 3 = 4$ and y-coordinate $2 + 1 = 3$.
So P is at $(4, 3)$.

Length of a line segment

The length of AB in the diagram above can be found by Pythagoras' theorem.

$AC = 7 - 1 = 6$ units
$BC = 4 - 2 = 2$ units

So $\quad AB^2 = 6^2 + 2^2$
$AB^2 = 36 + 4$
$AB^2 = 40$
$AB = \sqrt{40}$

Examples   A line segment AB goes from point A at $(-3, 5)$ to point B at $(1, 2)$.
    Calculate the midpoint of the line segment AB.

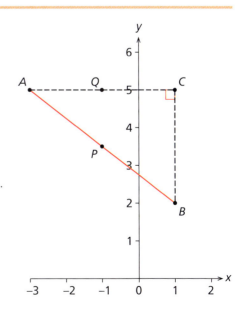

$AC = 4$ units so AQ is 2 units.
$CB = 3$ units so PQ is 1.5 units.
Therefore P has:
x-coordinate $-3 + 2 = -1$
and y-coordinate $5 - 1.5 = 3.5$.
So P is at $(-1, 3.5)$.

Calculate the length of the line segment AB.

Since $AC = 4$ units and $BC = 3$ units, using Pythagoras:

$(AB)^2 = 4^2 + 3^2$
$(AB)^2 = 16 + 9$
$(AB)^2 = 25$
So $AB = 5$ units.

Exercise 9.6

Use Pythagoras' theorem to calculate the length of the side marked ? in each of the following.

1 a

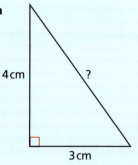

4 cm
?
3 cm

b

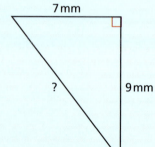

7 mm
?
9 mm

c

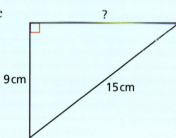

?
9 cm
15 cm

d

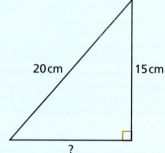

20 cm
15 cm
?

2 a

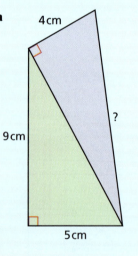

4 cm
?
9 cm
5 cm

b

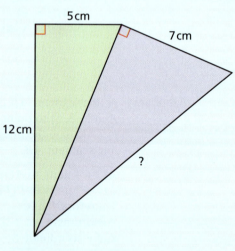

5 cm
7 cm
12 cm
?

c

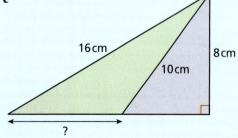

16 cm
10 cm
8 cm
?

d

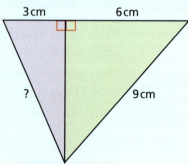

3 cm
6 cm
?
9 cm

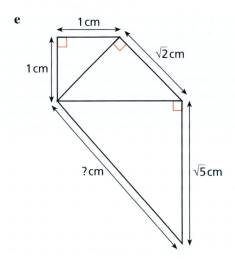

e

1 cm

1 cm

√2 cm

? cm

√5 cm

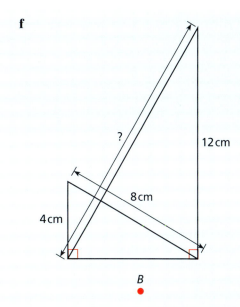

f

?

12 cm

8 cm

4 cm

3 Villages *A*, *B* and *C* lie on the edge of the Namib desert. Village *A* is 30 km due north of village *C*. Village *B* is 65 km due east of *A*. Calculate the shortest distance between villages *C* and *B*, giving your answer to the nearest 0.1 km.

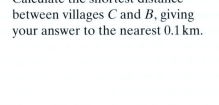

A

B

C

4 Town *X* is 54 km due west of town *Y*. The shortest distance between town *Y* and town *Z* is 86 km. If town *Z* is due south of *X*, calculate the distance between *X* and *Z*, giving your answer to the nearest 1 km.

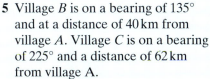

X

Y

Z

5 Village *B* is on a bearing of 135° and at a distance of 40 km from village *A*. Village *C* is on a bearing of 225° and a distance of 62 km from village A.
  a Show that triangle *ABC* is right-angled.
  b Calculate the distance from *B* to *C*, giving your answer to the nearest 0.1 km.

N

A

B

C

6 Two boats set off from X at the same time. Boat A sets off on a bearing of 325° and with a velocity of 14 km/h. Boat B sets off on a bearing of 235° with a velocity of 18 km/h.

Calculate the distance between both boats after they have been travelling for 2.5 hours. Give your answer to the nearest metre.

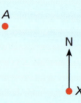

7 A boat sets off on a trip from Z. It heads towards B, a point 6 km away and due north. At B it changes direction and heads towards point C, also 6 km away and due east of B. At C it changes direction once again and heads on a bearing of 135° towards D, which is 13 km from C.

a Calculate the distance between Z and C to the nearest 0.1 km.

b Calculate the distance the boat will have to travel if it is to return to Z from D.

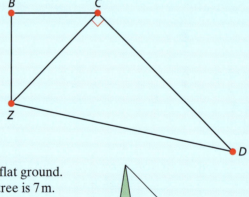

8 Two trees are standing on flat ground. The height of the smaller tree is 7 m. The distance between the top of the smaller tree and the base of the taller tree is 15 m. The distance between the top of the taller tree and the base of the smaller tree is 20 m.

a Calculate the distance between the two trees.

b Calculate the height of the taller tree.

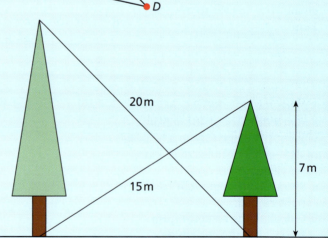

9 A line segment AB goes from the point A at $(-5, -1)$ to the point B at $(7, 4)$.

   a Calculate the coordinates of the mid-point of the line segment AB.

   b Calculate the length of AB.

10 A line segment AB goes from the point A at $(-2, 2)$ to the point B at $(4, -4)$.

   a Calculate the mid-point of the line segment.

   b Calculate the length of AB.

11 A line segment AB goes from point A at $(-5, -3)$ to point B at $(6, -3)$.

   a Calculate the mid-point of the line segment.

   b Calculate the length of AB.

12 A line segment AB goes from point A at $(3, 6)$ to point B at $(5, -2)$.

   a Calculate the mid-point of the line segment.

   b Calculate the length of AB.

> **Remember:**
> *To find the coordinates of the midpoint of a line segment AB add the x-coordinates and ÷2, add the y-coordinates and ÷2.*

Exercise 9.7

Use Pythagoras' theorem or trigonometry, or both, to answer the following questions. In each case give your answer to 1 d.p.

> **Remember:**
> *Always write a bearing with three digits.*

1 A sailing boat sets off from a point X and heads towards Y, a point 17 km north. At point Y, it changes direction and heads towards point Z, a point 12 km away on a bearing of 090°. When at Z, the boat wants to return to X. Calculate:

   a the distance ZX

   b the bearing of X from Z.

2 An aeroplane sets off from D on a bearing of 024° towards E, a point 250 km away. At E it changes course and heads towards F on a bearing of 055° and a distance of 180 km away.

   a How far is E to the north of D?

   b How far is E to the east of D?

   c How far is F to the north of E?

   d How far is F to the east of E?

   e Calculate the shortest distance between D and F.

   f Calculate the bearing of D from F.

3 Two trees are standing on flat ground. The angle of elevation of their tops from a point X on the ground is 40°. If the horizontal distance between X and the small tree is 8 m and the distance between the tops of the two trees is 20 m, calculate:
a the height of the small tree
b the height of the tall tree
c the horizontal distance between the trees.

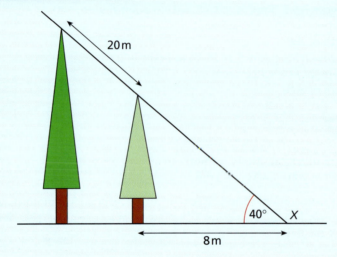

4 $PQRS$ is a quadrilateral. The sides RS and QR are the same length. The sides PQ and RS are parallel. Calculate:
a angle SQR
b angle PSQ
c length PQ
d length PS
e the area of $PQRS$.

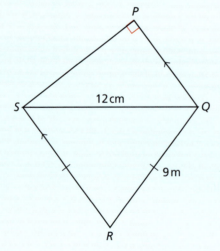

SUMMARY

By the time you have completed this chapter you should be able to:

■ solve problems involving **three-figure bearings** and **back bearings**
■ understand and use the trigonometric ratios **sine**, **cosine** and **tangent** in right-angled triangles, in two-dimensional problems.

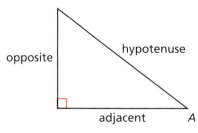

$$\sin A = \frac{\text{opposite}}{\text{hypotenuse}}$$

$$\cos A = \frac{\text{adjacent}}{\text{hypotenuse}}$$

$$\tan A = \frac{\text{opposite}}{\text{adjacent}}$$

The ratios can be remembered by the 'words' SOHCAHTOA. Sin is Opp over Hyp, Cos is Adj over Hyp, Tan is Opp over Adj.

■ understand, recall and use Pythagoras' theorem in two-dimensional problems, i.e. for any right-angled triangle, $a^2 = b^2 + c^2$ where a is the hypotenuse.

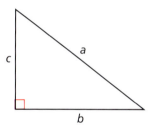

■ solve problems involving Pythagoras' theorem, trigonometry and bearings.

Exercise 9A

You will need:
• ruler
• protractor

1 From a village church in Millford it is possible to see the church spires of five neighbouring churches. The bearing and distance of each one from Millford church is given in the table below.

village	distance	bearing
Bourn	8 km	070°
Comberton	12 km	135°
Duxford	9 km	185°
Eversden	7.5 km	250°
Foxton	11 km	310°

a Choose an appropriate scale and draw the position of each church on a map.
b Using your map, work out the distance and bearing of the following:
  i) Bourn from Duxford
  ii) Eversden from Comberton.

2 A coastal radar station picks up a distress call from a ship. It is 50 km away on a bearing of 345°. The radar station contacts a lifeboat at sea which is 20 km away from the station on a bearing of 220°.
Make a scale drawing and use it to find the distance and bearing of the ship from the lifeboat.

3 Calculate the length of the side marked with a letter in each of the diagrams below.

a

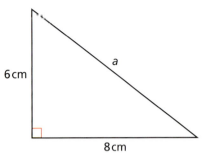

b

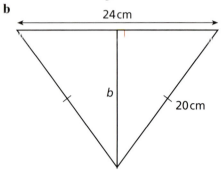

4 A rectangular swimming pool measures 50 m long by 15 m wide. Calculate the length of the diagonal of the pool. Give your answer correct to 1 d.p.

Exercise 9B

1 A map shows three towns *A*, *B* and *C*. Town *A* is due north of *C*. Town *B* is due east of *A*. The distance *AC* is 75 km and the bearing of *C* from *B* is 245°. Calculate, giving your answers to the nearest 100 m:
a the distance *AB*
b the distance *BC*.

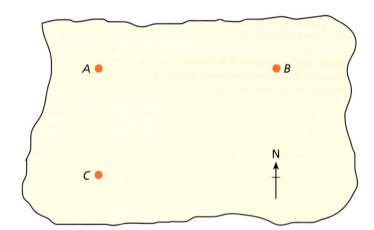

2 Two trees stand 16 m apart. The angle their tops make with a point *A* on the ground is *a* (in degrees).
a Express *a* in terms of the height of the shorter tree and its distance *x* from point *A*.
b Express *a* in terms of the height of the taller tree and its distance from *A*.
c Form an equation in terms of *x*.
d Calculate the value of *x*.
e Calculate the angle *a*.

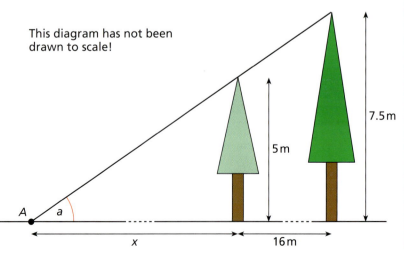

3 Two boats X and Y, sailing in a race, are shown in the diagram. Boat X is 145 m due north of a buoy B. Boat Y is due east of B. Boats X and Y are 320 m apart. Calculate:
   a the distance BY
   b the bearing of Y from X
   c the bearing of X from Y.

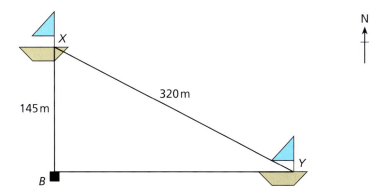

4 Two hawks P and Q are flying vertically above one another at a distance of 250 m apart. They both spot a snake at R.
   Using the information given, calculate:
   a the height of P above the ground
   b the distance between P and R
   c the distance between Q and R.

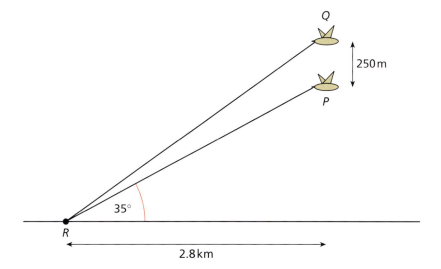

Exercise 9C

- Go outside and find a tall object (e.g. either a building or tree).
- Measure the horizontal distance *d* between you and the object.
- Measure the angle of elevation *a* from you to the top of the object (see the diagram).

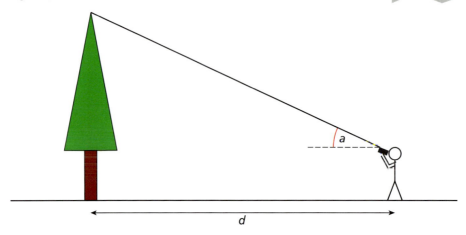

You will need:
- tape measure
- clinometer (to measure angles)
- calculator

- Repeat the process several times for different values of *d*.
- Record your results in a table as shown below.

object	result	distance *d*	angle *a*	height of object	average height
tree	1				
	2				
	3				
	4				

- Using trigonometry, use your values of *d* and *a* to calculate the height of the object.
- Why do you need to add your own height to the final result to get a true value for the height of the object?

Exercise 9D

You have seen in this chapter that Pythagoras' theorem states that, for any right-angled triangle, the area of the square on the hypotenuse is equal to the sum of the areas of the squares drawn on the other two sides.

For this exercise use a geometry package such as Cabri to investigate whether Pythagoras' theorem also works for semi-circles drawn on each of the sides of a right-angled triangle.

You will need:
- computer with Cabri or similar software installed

The picture below shows a right-angled triangle, with circles drawn on each side. The centre of each circle lies on the midpoint of each side. The diameter of the circle is equal to the length of the side on which it is constructed.

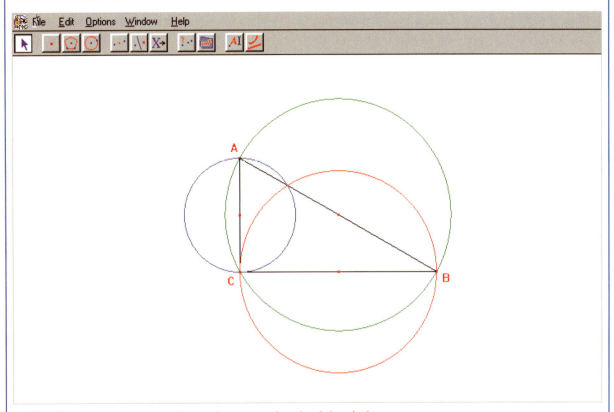

- Use the programme to work out the areas of each of the circles.
- Therefore work out the area of each of the semicircles drawn on each side.
- Does Pythagoras' rule still hold?
- Repeat the exercise with other right-angled triangles. Does Pythagoras' still work?

Exercise 9E

The Greek mathematician and philosopher Pythagoras formed a group of students who discovered important rules about harmony in music. Using the internet as a resource, find out more about these rules.

You will need:
- computer with internet access

10 Geometrical properties and representations

Geometry is a branch of mathematics concerned with the properties of shape and space. The geometry studied in this chapter is Euclidean geometry, after the Greek mathematician Euclid (about 300 BC). This is based on a few simple observations, such as: parallel lines never meet, and the interior angles of a triangle add up to 180°. There is a branch of geometry called **non-Euclidean geometry**, where these observations do not hold, for example on the surface of a **torus** (a doughnut shape). Much advanced mathematics is concerned with the study of non-Euclidean geometry.

Most buildings consciously or unconsciously echo the ideas of Euclidean geometry. For a start they 'look right'. However, some buildings of the twentieth century extended architectural ideas.

Le Corbusier (1887–1965) was as much artist as architect. His buildings had an enormous influence on the way other buildings were designed.

Euclid of Alexandria lived from about 365 BC to about 300 BC. He was the most prominent mathematician of ancient times, best known for his book on geometry, *The Elements*. Euclid's writings have been studied and used since then – over 2300 years!

Another example is the Sydney Opera House, designed by Jøern Utzon.

Similar triangles

Look at the two triangles below.

> **Remember:**
> *After an enlargement, a figure has the same shape but is a different size.*

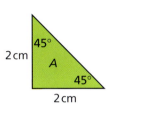
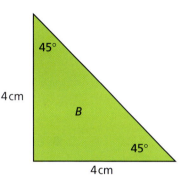

The lengths of the sides of triangle B are twice those of triangle A. The sizes of all three angles in B are the same as those in A. Therefore triangle B is an **enlargement** of triangle A. The ratio by which the lengths are changed is the **scale factor**.

If two triangles are **equi-angular** (have the same sized angles), but one is an enlargement of the other, then the triangles are said to be **similar**.

Note. If the 'enlarged' shape is smaller than the original, then the scale factor of enlargement is a figure between 0 and 1. In this diagram, triangle B is an enlargement of triangle A with scale factor $\frac{1}{2}$.

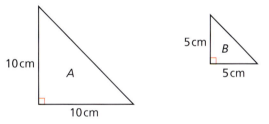

Example The isosceles triangles below are similar:

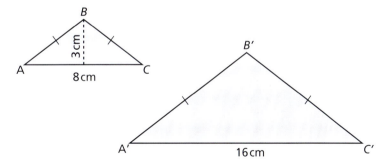

a Calculate the scale factor of enlargement.
b Calculate the height of $\triangle A'B'C'$.

a As $\triangle ABC$ and $\triangle A'B'C'$ are similar, the scale factor of enlargement can be calculated from the ratio of the lengths of **corresponding** sides. Sides AC and $A'C'$ are corresponding as they represent the same side on each diagram.

Therefore the scale factor of enlargement is $\frac{16}{8}$, i.e. 2.
b As the height of $\triangle ABC$ is 3cm, the height of $\triangle A'B'C'$ is 6cm.

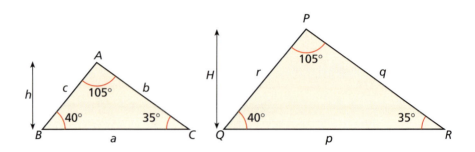

• *In general, therefore, if △ABC and △PQR are similar, as in the diagram, the ratios of the lengths of corresponding sides are the same and represent the scale factor, i.e.*

$$\frac{p}{a} = \frac{q}{b} = \frac{r}{c} = k \quad (\text{where } k \text{ is the scale factor of enlargement})$$

The heights of similar triangles are also proportional.

$$\frac{H}{h} = \frac{p}{a} = \frac{q}{b} = \frac{r}{c} = k$$

Exercise 10.1

1 a Explain why the two triangles below are similar.

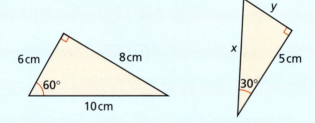

b Calculate the scale factor which enlarges the smaller triangle to the larger one.
c Calculate the values of x and y.
2 Which of the triangles below are similar?

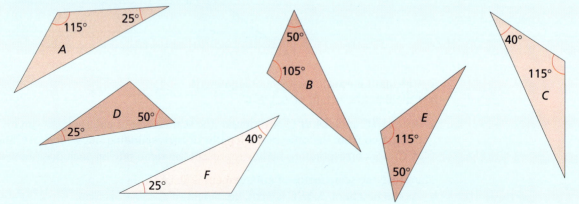

3 The triangles below are similar.

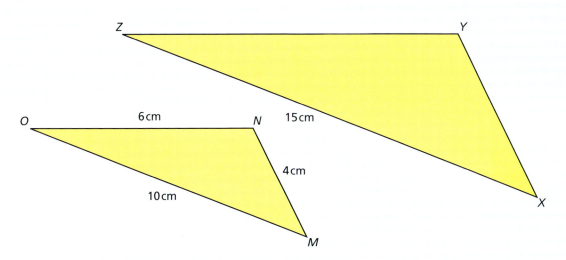

By finding the scale factor:
a calculate the length *XY*
b calculate the length *YZ*.

4 In the triangle below, calculate the lengths of *p*, *q* and *r*.

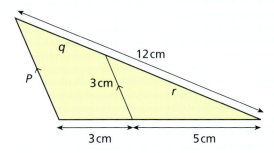

Hint: split the diagram into two similar triangles.

5 In the triangle below calculate the lengths of *e* and *f*.

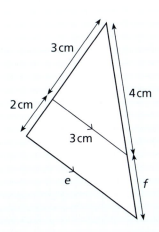

6 The triangles *ABC* and *XYZ* below are similar.

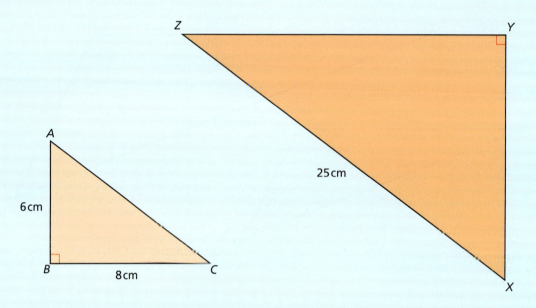

  a Using Pythagoras' theorem calculate the length *AC*.
  b Calculate the scale factor of enlargement.
  c Calculate the perimeter of $\triangle XYZ$.

7 The triangle *ADE* shown has an area of $12\,\text{cm}^2$.
  a Calculate the scale factor of enlargement from $\triangle ADE$ to $\triangle ABC$.
  b Calculate the length *DE*.
  c Calculate the length *BC*.

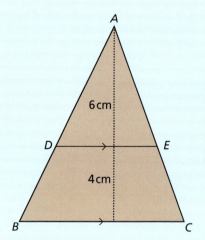

Exercise 10.2

1 In the regular hexagons below, hexagon *B* is an enlargement of hexagon *A* by a scale factor of 2.5.

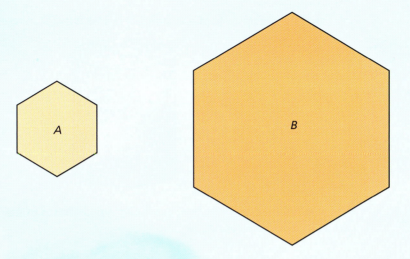

  a If the length of one of the sides of hexagon *A* is 4 cm, calculate the perimeter of hexagon *B*.
  b Each of the interior angles of hexagon *A* are 120°. What is the size of each of the interior angles of hexagon *B*?

2 *P* and *Q* are two regular pentagons. *Q* is an enlargement of *P* by a scale factor of 3. If the perimeter of pentagon *Q* is 90 cm, calculate the perimeter of *P*.

3 Below is a row of four triangles *A*, *B*, *C* and *D*. Each is an enlargement of the previous one by a scale factor of 1.5.

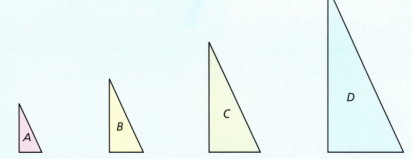

  a If the perimeter of *C* is 45 cm, calculate the perimeter of:
    i) triangle *D*
    ii) triangle *B*
    iii) triangle *A*
  b If the triangles were to continue in this sequence, which triangle would be the first to have a perimeter greater than 500 cm?

4 The sides of a square are increased by 10%. If the lengths of the sides of the enlarged square are 11 cm, calculate the length of each side of the original square.

5 The sides of a square of side length 4 cm are increased by 25% and then by a further 50%. Calculate the perimeter of the final square.

Geometrical views

All views of objects are either two-dimensional or three-dimensional. However, it is possible to draw a three-dimensional (3D) object using two-dimensional (2D) techniques.

The diagram below is of a cylinder; it is drawn in 2D to represent 3D.

> All 2D drawings of 3D solids involve some distortion. In this diagram, the circular top of the cylinder looks like an oval or an ellipse.

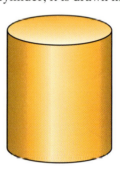

The diagrams below show how a cylinder can be drawn using **elevations**. These are 2D views seen from different angles.

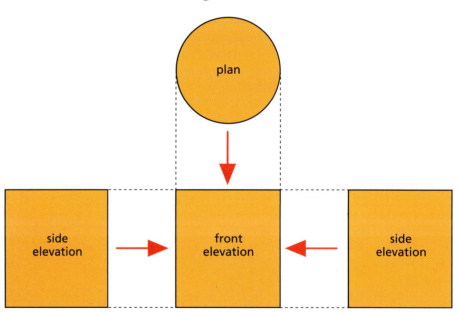

As the name suggests, the front elevation is the view seen from the front, while the side elevations are the views seen from the left and right sides. The top elevation is more commonly known as the **plan view**.

Note. The dimensions of each of the views must be consistent with each other. This is shown in the diagram above by the broken lines.

Exercise 10.3

For each of the 3D shapes below, draw front, side and plan elevations.

1

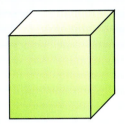

2

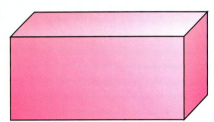

3

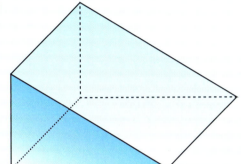

4

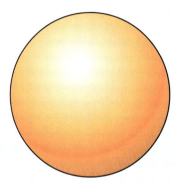

Exercise 10.4

In the following, sketch the 3D shape from the elevations given.

1

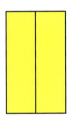

 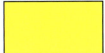

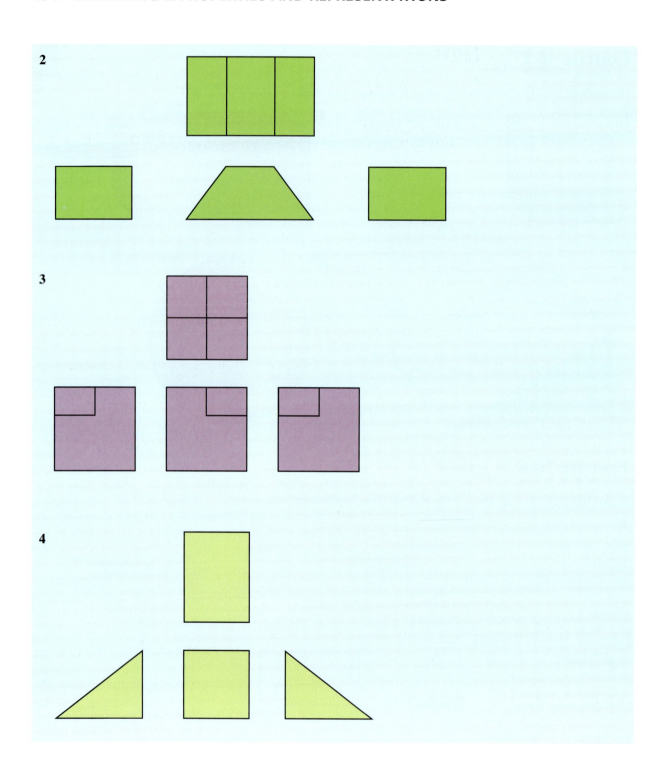

Nets

Nets, as you will already be aware, are a 2D representation of the faces of a 3D object. By folding the net up, the 3D object is created.

A knowledge and understanding of nets is essential to package designers, as most types of packaging start off as a net.

Below is the common net of a cube, which you will already be familiar with.

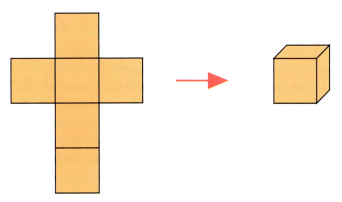

> There are at least ten possible nets of a cube.

This is, however, not the only net which will produce a cube. Below are two more examples.

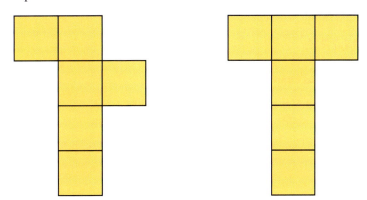

Exercise 10.5

For each of the following 3D shapes draw at least two different nets.

1

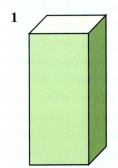

2

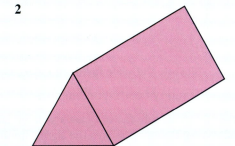

3

4

5

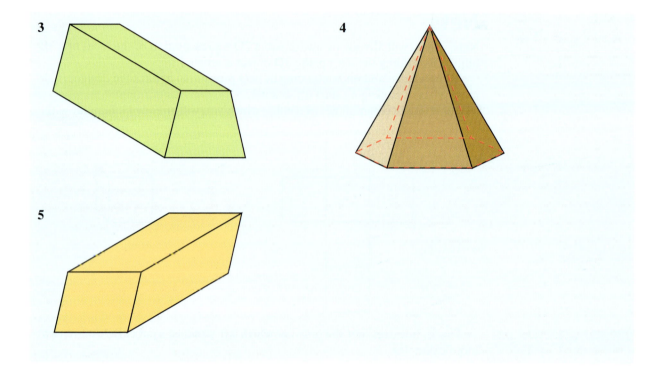

SUMMARY

By the end of this chapter you should:

■ understand what is meant by the term **similar**
■ know how to identify whether shapes are similar, i.e.

> the ratio of corresponding sides is constant in similar shapes
> angles are the same in similar shapes

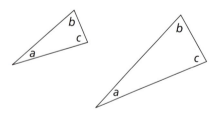

■ know how to calculate the **scale factor** of **enlargement** between an object and its image, i.e.

> calculate the ratio of corresponding sides

(Remember that the scale factor of enlargement can also be between 0 and 1. It is still called an enlargement even though it makes the figure smaller.)
■ know that 3D objects can be represented in 2D drawings by using front, side and plan **elevations**
■ be aware that a 3D object can have more than one net.

Exercise 10A

1 Which of the triangles below are similar?

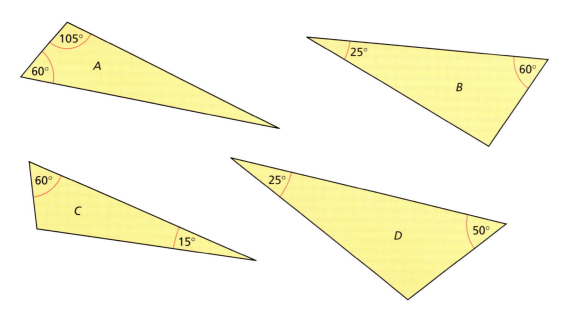

2 Using the triangle shown,
  a explain whether $\triangle ABC$ and $\triangle QBP$ are similar
  b calculate the length QB
  c calculate the length BC
  d calculate the length AP.

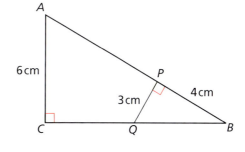

3 Two circles A and B are shown. Circle A has an area of $9\pi\,\text{cm}^2$, while circle B has an area of $36\pi\,\text{cm}^2$.
  a Calculate the radius of circle A.
  b Calculate the radius of circle B.
  c Calculate the scale factor of enlargement from A to B.

area = $9\pi\,\text{cm}^2$

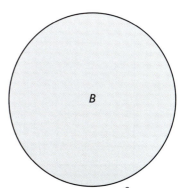

area = $36\pi\,\text{cm}^2$

4 Sketch the object which is represented by the elevations shown below.

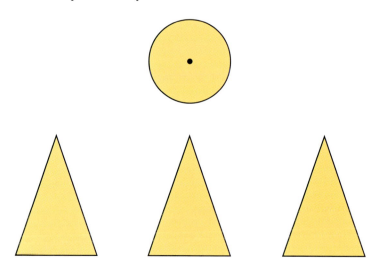

5 Which of the nets 1–4 below could be folded to make a triangular prism?

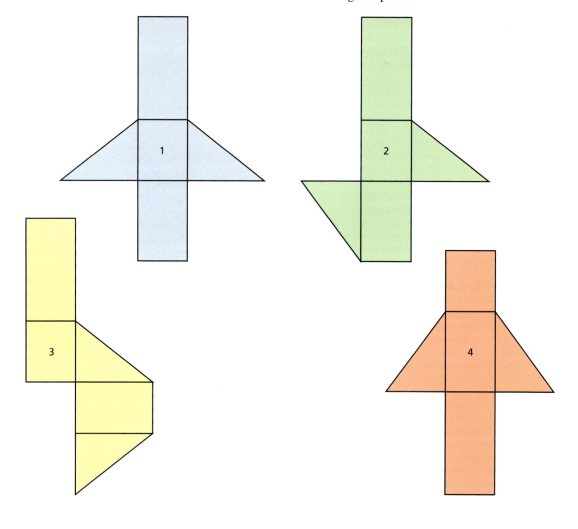

Exercise 10B

Diagrams are not to scale.

1 The two triangles below are similar.

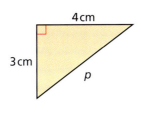
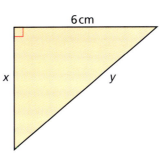

   a Using Pythagoras' theorem, calculate the length p.
   b Calculate the lengths x and y.

2 Calculate the lengths of x, y and z.

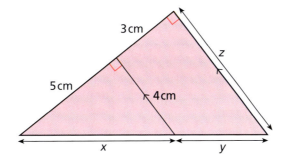

3 Rectangle B is an enlargement of rectangle A by a scale factor of 2.5. Use this information to calculate the dimensions of the sides marked x, y and z.

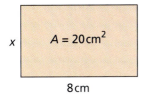

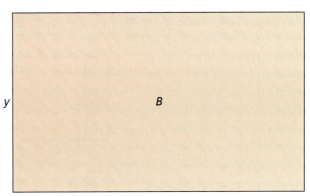

4 Which of the nets 1–4 could be folded to make a cuboid?

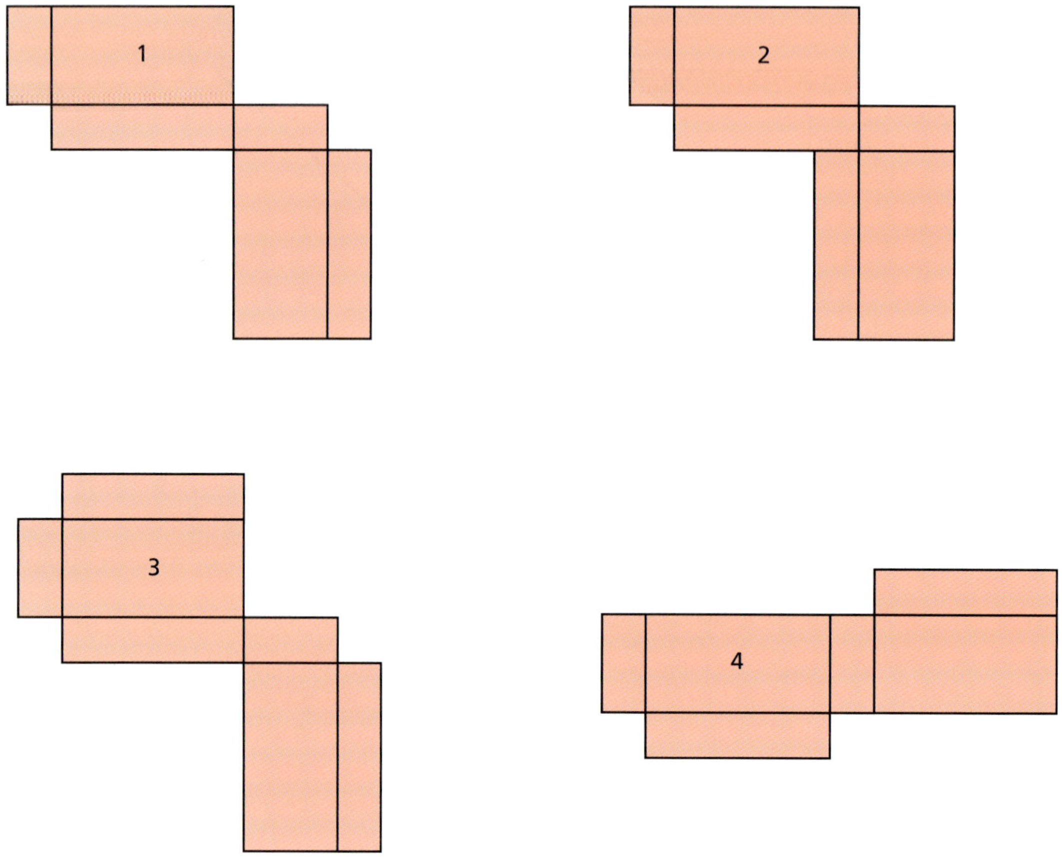

5 Sketch the object which is represented by the elevations shown below.

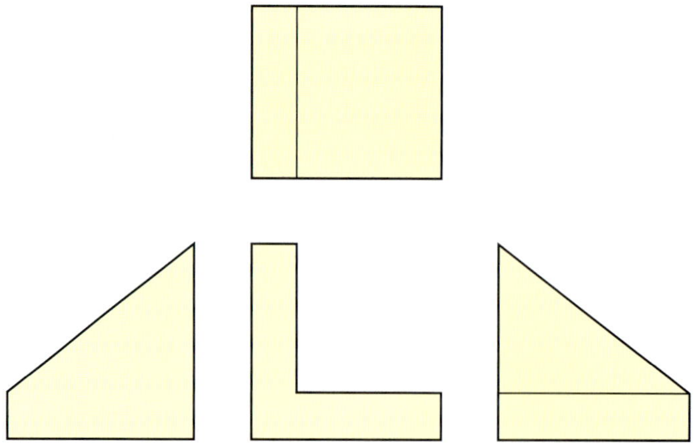

Exercise 10C

Eight equilateral triangles can be arranged as shown below.

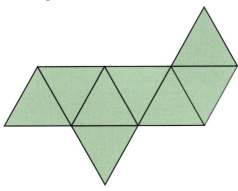

This arrangement can be folded to form an octahedron.

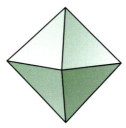

Draw other arrangements of eight equilateral triangles which also form octahedrons when folded.

Exercise 10D

From an A4 piece of card (21 cm × 29.5 cm) it is possible to construct an open-topped box. This is done by cutting shapes from each of the four corners as shown below.

You will need:
• computer with spreadsheet package installed

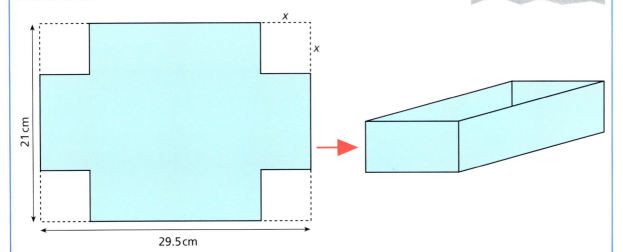

In the diagram the length of the sides of the squares is labelled *x*.

1 Calculate the volume of the box if *x* = 1 cm.
2 Calculate the volume of the box if *x* = 2 cm.

3 Design a spreadsheet to calculate the integer value of x that will produce the box with the greatest volume. You may wish to set your spreadsheet out in a similar way to the one shown below.

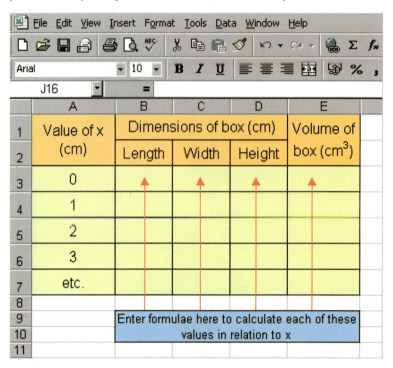

4 Plot a graph (with a smooth curve) to show how the volume of the box changes as x varies.
5 Does your graph show the maximum volume occurring at an integer value of x?
6 If your answer to question 5 is 'no', use your spreadsheet to calculate the value of x to 1 d.p. which gives the maximum volume.

Exercise 10E

Euclid made five principal statements, called the five postulates. Using the internet as a resource, find out what these five postulates stated. Find out also where non-Euclidean geometry differs from these postulates.

You will need:
• computer with internet access

11 Measures

In one day a soldier in the army of Julius Caesar could comfortably march 20 miles with full kit and then help to build a defensive blockade.

The mile was a unit of length based upon 1000 strides of a Roman Legionary. As such the measurement was sufficiently accurate for its purpose but only an approximate distance.

Most measures started as rough estimates. The yard was said to be the distance from the king's nose (alleged to be Henry I) to the tip of his extended finger. As it became necessary to have standardisation in measurement, the measures themselves became more exact.

In 1791 during the French Revolution, a new unit of measurement, the metre, was defined in France. Originally it was defined as 'one ten millionth of the length of the quadrant of the Earth's meridian through Paris'. The use of this unit of measurement became law in France in 1795.

However, this measurement was not considered sufficiently accurate and further definitions were required.

In 1927 the metre was defined as the distance between two marks on a platinum–iridium bar. This bar is kept in Paris.

In 1960 the definition was based on the emission of a krypton 86 lamp.

At the 1983 General Conference on Weights and Measures, the metre was re-defined as the length of the path travelled by light in a vacuum in $\frac{1}{299792483}$ second. This not very neat definition can be considered one of the few 'accurate' measures. Most measures are only to a degree of accuracy.

In this chapter we will look at the limits of accuracy when a number is given correct to the nearest whole number, or to a given number of decimal places.

Imperial measure

Two of the units of length mentioned above, the mile and the yard, are examples of **'Imperial' measures**, which are still used. Other Imperial measures that you, or certainly older members of your family, will be familiar with are pound (lb), stone and ton for weight, pint and gallon for capacity, and inch and foot for length. There were many more Imperial units, including rod, chain, furlong, gill, quart, ounce, hundredweight and many others.

Try to find out more about these.

Below are some of the more common Imperial measures and their metric equivalents.

> 1 kilogram is about 2.2 lb
> 8 kilometres is about 5 miles
> 1 metre is about 39 inches
> 1 litre is about 1.75 pints

Example   If 1 kg is approximately 2.2 lb, what is the approximate metric equivalent of 1 lb?

1 kg is approximately 2.2 lb.

So $\dfrac{1}{2.2}$ kg is approximately 1 lb.

Therefore the approximate metric equivalent of 1 lb is 0.45 kg or 450 g.

Exercise 11.1

(*Revision*)
Copy and complete the table below.

Imperial measure	approximate metric equivalent
1 lb (pound)	450 g
1 mile	____ m
1 inch	____ cm
1 foot (12 inches)	____ cm
1 yard (3 feet)	____ cm
1 pint	____ cl
1 gallon	____ l
1 ton (2240 lb)	____ kg
1 square yard	____ m²
1 acre (4840 sq yd)	____ m²

Accuracy

Numbers can be written to different **degrees of accuracy**. For example 4, 4.0 and 4.00 appear to represent the same number, but they may not. This is because they are written to different degrees of accuracy.

4 is rounded to 0 decimal places and therefore any number from 3.5 up to *but not including* 4.5 would be rounded to 4. On a number line this would be represented as:

> **Remember:**
> ● means the number is included
> ○ means it isn't included.

As an **inequality**, where x represents the number, it would be expressed as

$$3.5 \leqslant x < 4.5$$

In this example, 3.5 is known as the **lower bound** of 4, while 4.5 is known as the **upper bound**.

4.0 on the other hand is written to one decimal place and therefore only numbers from 3.95 up to but not including 4.05 would be rounded to 4.0. This therefore represents a much smaller range of numbers than the range of those being rounded to 4.

Similarly the range of numbers being rounded to 4.00 would be even smaller. This is shown on the number line below.

Example A girl's height is given as 162 cm to the nearest centimetre.
 a Work out the lower and upper bounds within which her height can lie.
 b Represent this range of numbers on a number line.
 c If the girl's height is h cm, express the range of possible values as an equality.

 a Lower bound = 161.5 cm.
  Upper bound = 162.5 cm.
 b

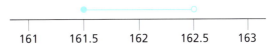

 c $161.5 \leqslant h < 162.5$

Exercise 11.2

1 Each of the numbers given below is expressed to the nearest whole number. For each number,
   i) give the upper and lower bounds
   ii) using x as the number, express as an inequality the range in which the number lies.
   a 6 **b** 83 **c** 152 **d** 1000 **e** 0

2 Each of the numbers given below is correct to one decimal place. For each number,
   i) give the upper and lower bounds
   ii) using x as the number, express as an inequality the range in which the number lies.
   a 3.8 **b** 15.6 **c** 1.0 **d** 10.0 **e** 0.3

3 Each of the numbers given below is correct to two significant figures. For each number,
   i) give the upper and lower bounds
   ii) using x as the number, express as an inequality the range in which the number lies.

   a 4.2 **b** 0.84 **c** 420

   d 5000 **e** 0.045 **f** 25 000

   > **Remember:**
   > The first non-zero digit of a number is its first significant figure.

4 The mass of a sack of vegetables is given as 5.4 kg.
   a Illustrate the lower and upper bounds of the mass on a number line.
   b Using M kg for the mass, express as an inequality the range of values in which M must lie.

5 At a school sports day, the winning time for the 100 m was given as 11.8 seconds.
   a Illustrate the lower and upper bounds of the time on a number line.
   b Using T seconds for the time, express as an inequality the range of values in which T must lie.

6 The capacity of a swimming pool is given as 620 m³ correct to two significant figures.
   a Calculate the lower and upper bounds of the pool's capacity.
   b Using x cubic metres for the capacity, express as an inequality the range of values in which x must lie.

7 A farmer measures the dimensions of his rectangular field to the nearest 10 m. The length is recorded as 630 m and the width is recorded as 400 m.
   a Calculate the lower and upper bounds of the length.
   b Using W metres for the width, express as an inequality the range of values in which W must lie.

Compound measures

Two examples of **compound measures** are:

    speed – measured in cm/s, m/s, miles per hour, km/h
    density – measured in g/cm³ (grams per cubic cm) or lb/cubic inch

A compound measure is one that is made up of two or more other measures. The formula for speed is:

$$\text{speed} = \frac{\text{distance}}{\text{time}}$$ sometimes represented as

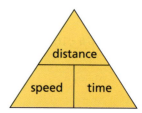

Therefore, to measure speed, we need to use two other types of measure, namely distance and time.

Exercise 11.3

(*Revision*)

1 Find the mean speed of an object moving:

    a 80 m in 5 s   **b** 72 cm in 6 s

    c 240 miles in 6 hours   **d** 600 km in 6 s

2 How far will an object travel in:

    a 5 s at 40 m/s   **b** 3 hours at 60 km/h

    c 2 hours 15 min at 60 m.p.h.   **d** 8 s at 12 cm/s

3 How long will an object take to travel:

    a 100 m at 8 m/s   **b** 2 km at 40 m/s

    c 15 miles at 60 m.p.h.   **d** 1 cm at 50 cm/s

Density is also a compound measure, defined as the mass of a substance in one unit of volume.

    It can be calculated from the formula

$$\text{density} = \frac{\text{mass}}{\text{volume}}$$

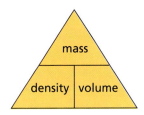

Examples   Find the density of a stone of mass 120 g and volume 20 cm³.

$$\text{density} = \frac{\text{mass}}{\text{volume}} = \frac{120}{20}$$

Therefore density of the stone = 6 g/cm³.

....

An object has a density of 25 g/cm³ and volume 120 cm³. Calculate the mass of the object.

$$\text{density} = \frac{\text{mass}}{\text{volume}}$$

Rearranging the formula gives:

$$\begin{aligned} \text{mass} &= \text{volume} \times \text{density} \\ &= 120 \times 25 \\ &= 3000 \end{aligned}$$

Therefore the object has mass 3000 g.

Exercise 11.4

1 Calculate the density of a substance of:
   a mass 2000 g   volume 40 cm^3
   b mass 4 kg   volume 125 m^3
   c mass 1 tonne   volume 2.5×10^6 cm^3 (give your answer in g/cm^3)
   d mass 2500 lb   volume 500 cubic inches
2 Calculate the mass of an object given that:
   a the density is 25 kg/m^3 and volume is 40 m^3
   b the density is 40 g/cm^3 and volume is 2000 cm^3
   c the density is 0.4 g/cm^3 and volume is 4.8×10^6 cm^3
   d the density is 14 lb/cubic inch and volume is 8 cubic inches
3 Calculate the volume of a substance of:
   a mass 500 g, density 25 g/cm^3
   b mass 15 g, density 8 g/cm^3
   c mass 2.5×10^8 g, density 0.5 g/cm^3
   d mass 700 lb, density 2.8 lb/cubic inch

SUMMARY

When you have completed this chapter you should know:

■ common **Imperial measures** and their metric equivalents
■ that sometimes a number is given to a certain **degree of accuracy** and that as a result it represents a range of numbers
■ that this range lies between the **upper bound** and **lower bound** and can be shown on a number line, for example 27 to the nearest whole number is shown as

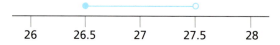

| 26 | 26.5 | 27 | 27.5 | 28 |

■ that the range for a number given to a degree of accuracy can be written as an **inequality**, for example
    $26.5 \leqslant x < 27.5$

■ what is meant by a compound measure, and give examples, such as speed and **density**.

Exercise 11A

1 The following numbers are expressed to the nearest whole number. Illustrate on a number line the range in which each must lie.
   a 7   b 40   c 0
2 The following numbers are expressed correct to two significant figures. Representing each number by the letter x, express the range in which it must lie, by using an inequality.
   a 210   b 64   c 3.0   d 0.88
3 A school janitor measures the dimensions of the rectangular playing field to the nearest metre. The length is recorded as 350 m and the width as 200 m. Express the range in which the length and width lie by using inequalities.

4 A boy's mass was measured to the nearest 0.1 kg. If his mass was recorded as 58.9 kg, illustrate on a number line the range within which his mass must lie.

5 An electronic clock is accurate to $\frac{1}{1000}$ of a second. The duration of a flash from a camera is timed on the clock at 0.004 seconds. Express the upper and lower bounds of the duration of the flash by using inequalities.

6 The following numbers are rounded to the degree of accuracy shown in brackets. Express the lower and upper bounds of each of these numbers as an inequality.

    a $x = 4.83$ (2 d.p.)   **b** $y = 5.05$ (2 d.p.)   **c** $z = 10.0$ (1 d.p.)

7 Calculate the density of a gas of mass 5.5 g and volume 1.1×10^5 cm^3.

Exercise 11B

1 The following numbers are rounded to the nearest 100. Illustrate on a number line the range in which each must lie.

    a 500   **b** 7000   **c** 0

2 The following numbers are expressed correct to three significant figures. Represent the limits of each number by using inequalities.

    a 254   **b** 40.5   **c** 0.410   **d** 100

3 The dimensions of a rectangular courtyard are given to the nearest 0.5 m. The length is recorded as 20.5 m and the width as 10.0 m. Represent the limits of these dimensions by using inequalities.

4 The circumference c of a tree is measured to the nearest 2 mm. If its circumference is measured as 245.6 cm, illustrate on a number line the range within which it must lie.

5 The time it takes Earth to rotate around the Sun is given as 365.25 days correct to five significant figures. What are the upper and lower bounds of this time?

6 The following numbers are rounded to the degree of accuracy shown in brackets. Express the lower and upper bounds of each of these numbers as an inequality.

    a 10.90 (2 d.p.)   **b** 3.00 (2 d.p.)   **c** 0.5 (1 d.p.)

7 Calculate the volume of a gas of mass 2.5 g and density 1.25×10^{-4} g/cm^3.

Exercise 11C Ma1

You will need:
• computer with spreadsheet package installed

A farmer has purchased a number of fencing panels, each 1 m in length. He wishes to make a rectangular enclosure, but is not sure of the optimum arrangement of panels to maximise the area.

With 100 panels, he can make a variety of different-sized enclosures, for example

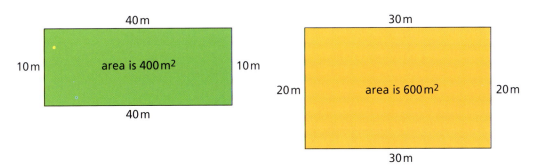

• Using a spreadsheet, investigate the greatest area possible with 100 m of fencing.
• Investigate the maximum area possible for fencing of length X m.
• Investigate the effect on the maximum area if each of the fencing panels were 1 m in length correct to the nearest centimetre.

Exercise 11D

The graph below was obtained using a graphical calculator linked to a motion sensor. The result is a distance–time graph.

- Explain what the gradient of a distance–time graph shows.
- Describe the movements that would have produced a distance–time graph similar to the one below.

You will need:
- graphical calculator
- motion sensor

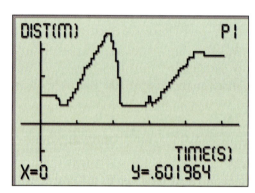

- Using a motion sensor produce some more distance–time graphs and compare them with their velocity–time graphs.

Exercise 11E

Robert Boyle (1627–1691) worked with Robert Hook (1653–1703) on gases and vacuums. Jacques Charles (1746–1823) later worked on the laws governing the expansion of gases. All three have laws of physics named after them.

Using an encyclopaedia or the internet as a resource, define and explain these laws.

You will need:
- encyclopaedia (book or CD rom) or
- computer with internet access

12 Geometry of the circle

The application of the properties of the circle through the invention of the wheel was of fundamental importance in the development of human society. The first 'wheels' were probably rollers made from trunks of trees which enabled large weights to be pulled along. The axle and solid wheel, still seen in carts used in some less developed parts of the world, was the next modification, which led to a revolution in transport.

The development of the water wheel was a major technological breakthrough, as was the circular mariner's compass and circular cog-wheels in gears.

Vocabulary of the circle

This introductory section explains some of the basic vocabulary in this chapter. It is assumed that you already know terms like diameter, arc, circumference and tangent.

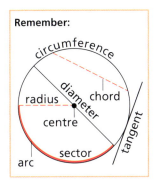
Subtend

In the circle shown, the arc *PQ* **subtends** (is opposite) an angle at *R*.

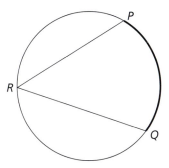

Segment

The shaded area in the diagram below is known as a **segment**. It is the area between a chord and an arc.

The **unshaded** area is also a segment.

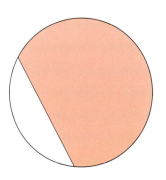

Concyclic points and a cyclic quadrilateral

In the diagram, the points *P*, *Q*, *R* and *S* are on the circumference of the circle and are known as **concyclic points**. Joining them produces the **cyclic quadrilateral** *PQRS*.

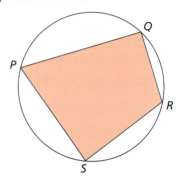

Angle at the centre of a circle

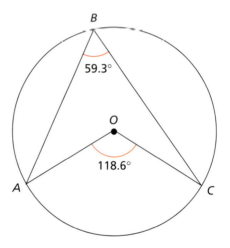

The diagram above shows a quadrilateral drawn inside a circle in such a way that one of its vertices is at the centre of the circle and the other three vertices lie on the circumference. The sizes of two angles are also given. Note that both angles are **subtended** by the same arc.

The diagram shows that one angle is double the size of the other. This can be proved always to be the case, as shown below.

Remember:
In triangle ABC, corners A, B and C are known as **vertices**.

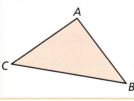

'Subtended by' means 'is opposite to'.

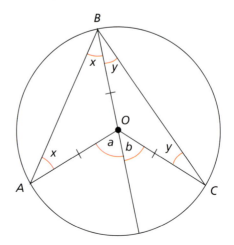

Triangles *AOB* and *BOC* are both isosceles as the sides *OA*, *OB* and *OC* are all radii of the circle.

$a = 2x$ (exterior angle of a triangle is equal to the sum of the opposite interior angles)

Similarly

$b = 2y$

Therefore

$a + b = 2x + 2y$

i.e.

$a + b = 2(x + y)$

These findings can be written as the following rule:

● *The angle subtended at the centre of a circle by an arc is twice the size of any angle on the circumference subtended by the same arc.*

The diagrams below illustrate this theorem.

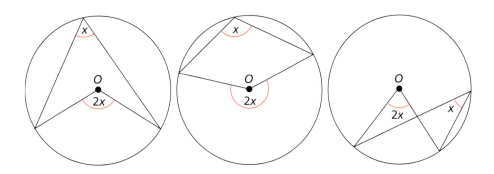

Exercise 12.1

Calculate the size of the angles marked *x* in each of the following diagrams.

1

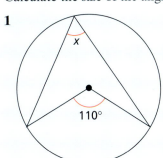

2

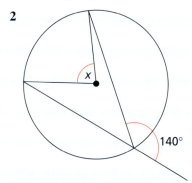

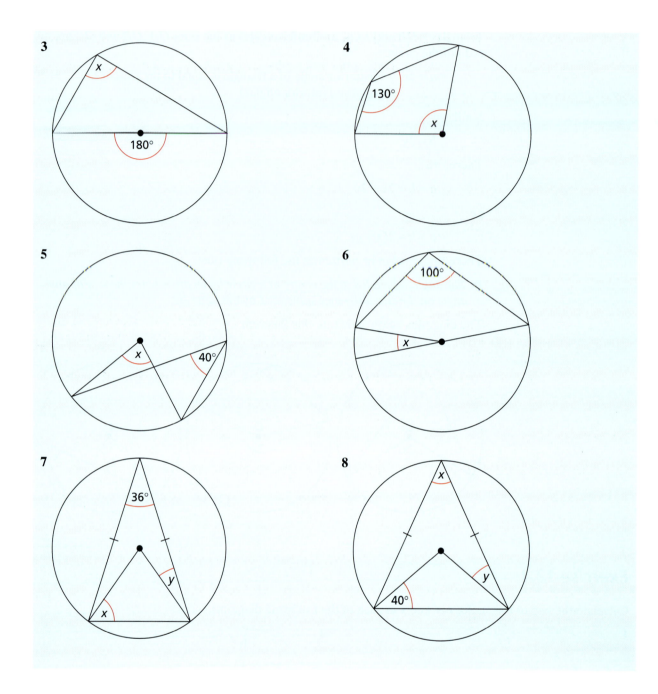

Angle in a semi-circle

The law for an angle at the centre of a circle can be extended to the situation described below.

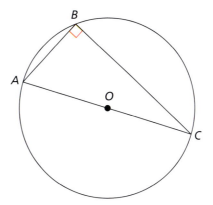

In the diagram above, $\angle AOC$ is 180° as it is formed by the diameter of the circle. This must be twice as large as the internal angle subtended by the same arc, i.e. $\angle ABC$. Therefore $\angle ABC$ is a right angle.

● *The angle subtended at the circumference by a semi-circle is a right angle.*

Exercise 12.2

Calculate the size of the angle marked x in each of the following diagrams.

1

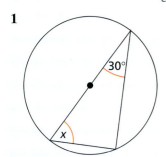

2

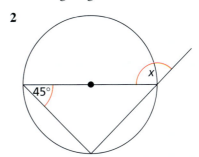

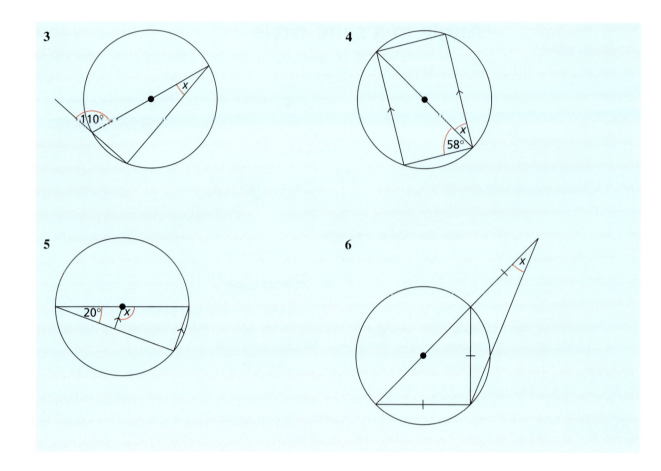

Angle between a tangent and a radius of a circle

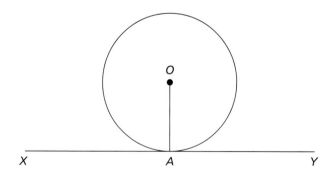

O is the centre of the circle, XY is a tangent meeting the circle at A, and OA is a radius. OA is a line of symmetry of the diagram, so

$$\angle OAX = \angle OAY = 90°$$

● *The angle between a tangent to a point on a circle and a radius at the same point on the circle is a right angle.*

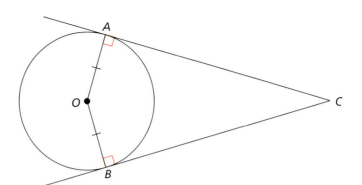

In $\triangle OAC$, $\angle OAC$ is a right angle. By applying Pythagoras' theorem to the triangle, it can be shown that

$$OC^2 = AC^2 + OA^2$$

$\triangle OAC$ and $\triangle OBC$ are also **congruent** as OAC and OBC are right angles. $OA = OB$ because they are both radii, and OC is common to both triangles. Therefore

$$AC = BC$$

● *The tangents from a point to a circle are equal in length.*

Exercise 12.3

In questions 1–6, calculate the size of the angle marked x in each diagram.

1

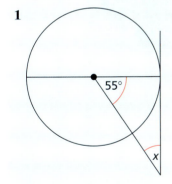

2

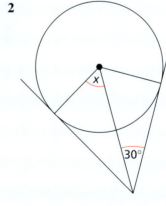

3

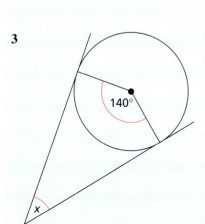

4

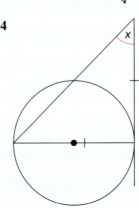

5

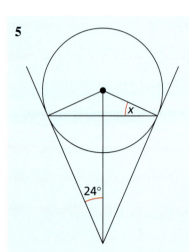

6

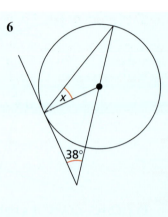

In questions 7–9, calculate the length of the side marked *x* in each diagram. (Use Pythagoras' theorem.)

7

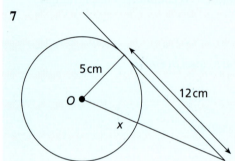

8

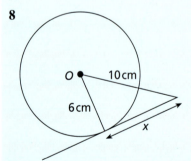

9

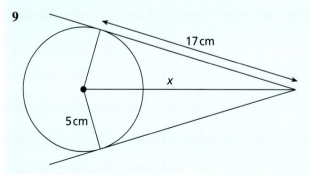

Angles in the same segment

Look at the diagram below. If the angle at the centre is $2x$, then each of the angles at the circumference must be equal to x. This can be explained by using the theorem that the angle subtended at the centre by an arc is twice any angle on the circumference subtended by the same arc.

All angles subtended at the circumference by a particular arc are angles in the same segment.

● *Angles in the same segment of a circle are equal.*

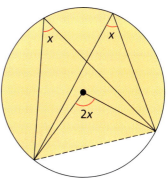

Exercise 12.4

Calculate the unknown angles in the following.

1

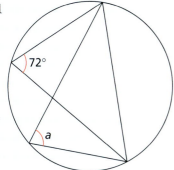

2

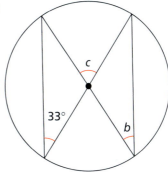

3

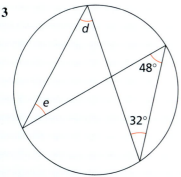

4

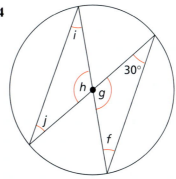

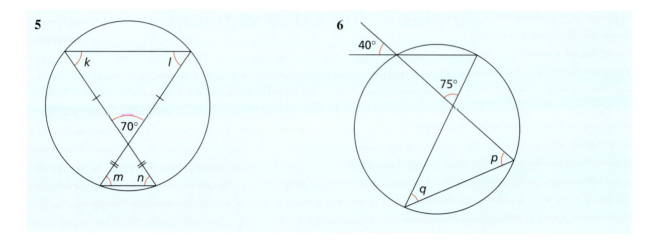

Angles in opposite segments

Points P, Q, R and S all lie on the circumference of the circle below, i.e. they are concyclic points. As we have seen, joining the points P, Q, R and S produces a cyclic quadrilateral.

● *The opposite angles in a cyclic quadrilateral are **supplementary**, i.e. they add up to 180°.*

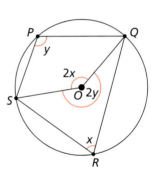

Since

$$p + r = 180° \quad \text{(supplementary angles)}$$

and

$$r + t = 180° \quad \text{(angles on a straight line)}$$

it follows that $p = t$.

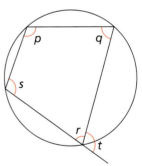

● *The exterior angle of a cyclic quadrilateral is equal to the interior opposite angle.*

Exercise 12.5

Calculate the size of the unknown angles in each of the following.

1

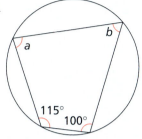

2

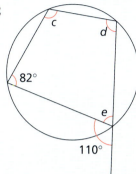

3

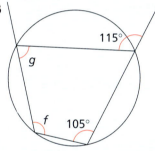

4

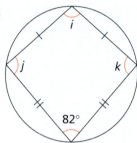

5

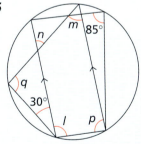

6

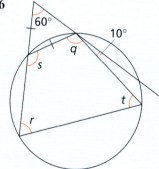

Perpendicular from the centre to a chord

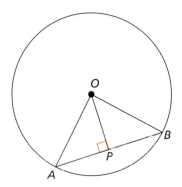

If O is the centre of the circle and AB is a chord, then OA and OB are radii.

If OP is perpendicular to AB then $\angle OPA$ and $\angle OPB$ are right angles. It follows that $\triangle AOP$ and $\triangle BOP$ are congruent, because:

- both triangles have a right angle ($\angle OPA$ and $\angle OPB$)
- the hypotenuses are of equal length (both are radii)
- both triangles have one side in common (side OP).

As the triangles are congruent then $AP = PB$; therefore OP, the perpendicular, bisects the chord AB.

● *The perpendicular from the centre of a circle to a chord bisects the chord.*

Exercise 12.6

(In this exercise, use Pythagoras' theorem and trigonometry to calculate the required lengths.)

In each of the questions, AB is a chord of a circle with centre O, and OP is a perpendicular to AB (as in the diagram above). Draw a sketch to help you.

1 If AB is 8 cm and the radius of the circle is 5 cm, calculate the length OP.

2 If $\angle AOB$ is 120° and AO is 5 cm, calculate the length of OP and of chord AB.

3 If $\angle OAP$ is 45° and AO is 5 cm, calculate the length of OP and of chord AB.

4 If chord AB is 24 cm long, calculate the radius of the circle if OP is 5 cm long.

5 Given a circle of centre O and radius r cm, show that if a chord AB is also r cm long, then the length of the perpendicular bisector OP of the chord can be expressed as $\dfrac{\sqrt{3}}{2}r$.

SUMMARY

By the end of this chapter you should know:

■ that the angle **subtended** at the centre of the circle by an arc is twice the angle subtended by the same arc at any point of the circumference

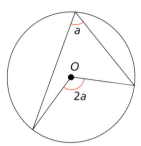

■ that the angle subtended at the circumference by a semi-circle is a right angle

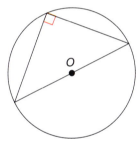

■ that the tangent at any point on a circle is perpendicular to the radius of the circle at that point

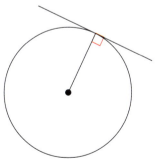

■ that tangents from a point to a circle are equal in length

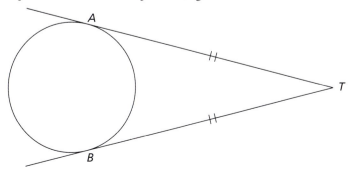

■ that angles in the same **segment** of a circle are equal

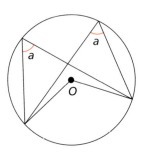

■ that opposite angles of a **cyclic quadrilateral** are **supplementary**

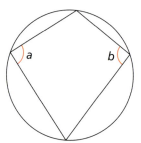

$a + b = 180°$

■ that the exterior angle of a cyclic quadrilateral is equal to the interior opposite angle

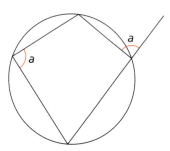

■ that the perpendicular from the centre of a circle to a chord bisects that chord.

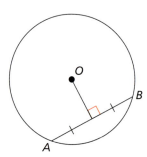

Exercise 12A

1 In each of the following diagrams, identify which angles are:
 i) supplementary angles
 ii) right angles
 iii) equal.

a

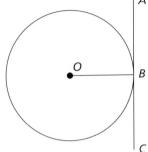

b

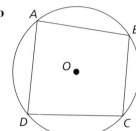

c

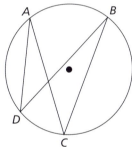

d

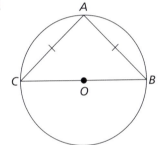

2 If $\angle POQ = 84°$ in this diagram, calculate
  a $\angle PRQ$ **b** $\angle OQR$

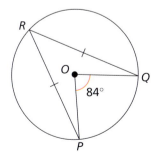

3 Calculate $\angle DAB$ and $\angle ABC$ in this diagram.

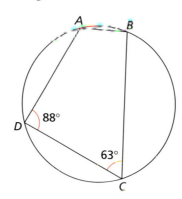

4 If DC is a diameter in this diagram, calculate angles BDC and DAB.

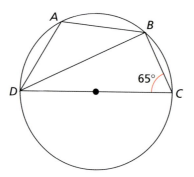

5 Calculate as many angles as possible in the diagram below. Explain your reasonings fully.

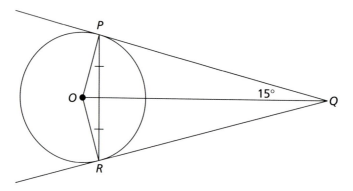

Exercise 12B

1 In each of the following diagrams, identify which angles are:
   i) supplementary angles
   ii) right angles
   iii) equal.

a

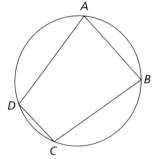

b

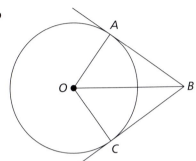

c

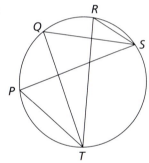

d

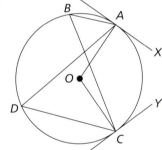

2 If $\angle AOC$ is $72°$ in this diagram, calculate $\angle ABC$.

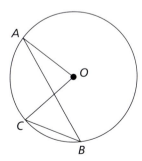

3 If $\angle AOB = 130°$ in this diagram, calculate $\angle ABC$, $\angle OAB$ and $\angle CAO$.

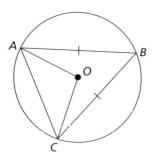

4 Show that $ABCD$ is a cyclic quadrilateral.

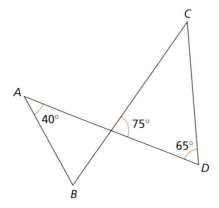

5 Calculate f and g.

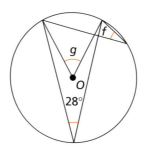

Exercise 12C

The diagram below shows two equilateral triangles ABC and ADE.

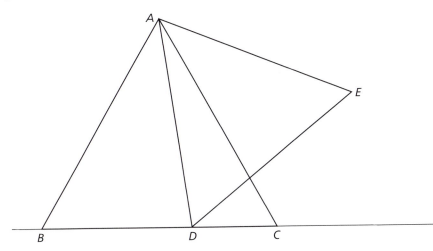

Vertex D is free to move to any position along the line BC and beyond as shown below.

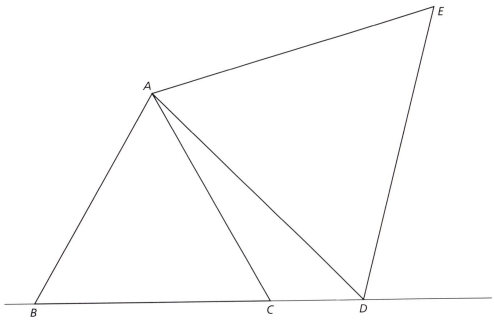

Note. Since triangle ADE remains equilateral, its size changes as D moves. Investigate the nature of the shape $ACDE$, giving full explanations to any conclusions that you make.

> An interactive version of this problem can be found at the following website: http://www.ies.co.jp/math/java/geo/inquad/inquad.html

Exercise 12D

Using Cabri II or a similar geometry package, model some of the circle theorems covered in this chapter.
Print them out and use these to produce a display on circle theorems.

You will need:
- computer with Cabri II or similar geometry package installed

Exercise 12E

In many countries, proof of theorems still forms much of the geometry studied in the secondary mathematics syllabus. This used to be true in the UK 30 years ago. See if you can find examples of proofs, possibly from the internet, or old textbooks.

13 Transformations

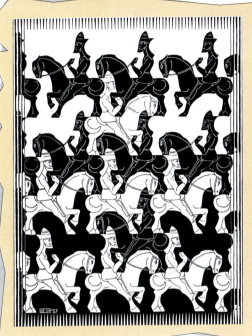

Maurits Escher was a famous twentieth-century Dutch artist. His work was striking because it combined art with mathematics. He was fascinated by the geometric patterns he saw in the tilings at the Alhambra palace in Granada, Spain. By studying their properties, he was able to take art and mathematics to new levels.

Much of his well-known work deals with **transformations**. By transforming polygons, Escher was able to create stunning tessellating patterns.

We saw in *Intermediate 1* that an object undergoing a transformation changes either its position or its shape. The transformations covered included **reflection** and **rotation**.

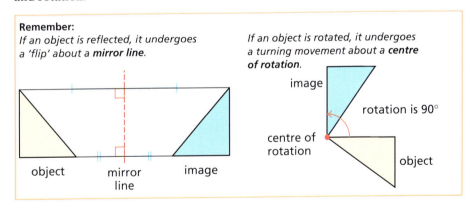

Remember:
*If an object is reflected, it undergoes a 'flip' about a **mirror line**.*

object   mirror line   image

*If an object is rotated, it undergoes a turning movement about a **centre of rotation**.*

image
rotation is 90°
centre of rotation
object

In this chapter we will deal with two further transformations, **enlargement** and **translation**.

Constructing enlargements

As we have already seen, if an object is enlarged, the result is an image which is mathematically similar to the object but of a different size. The image can be either larger or smaller than the original object.

When describing a transformation as an enlargement, two additional pieces of information need to be given: the position of the **centre of enlargement** and the **scale factor of enlargement**.

Examples   In the diagram below, triangle ABC is enlarged to form triangle $A'B'C'$.
a Find the centre of enlargement.
b Calculate the scale factor of enlargement.

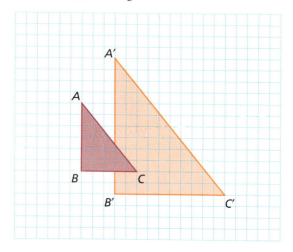

a To find the centre of enlargement, join corresponding points on the object and image with a straight line. Then extend these lines until they meet. The point at which they meet is the centre of enlargement O.

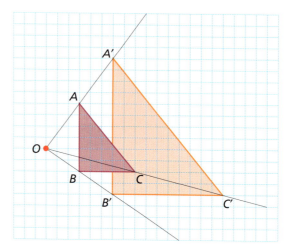

b You can calculate the scale factor of enlargement in one of two ways. From the diagram above, you can see that the distance OA' is twice the distance OA. Similarly OC' and OB' are both twice OC and OB respectively. Hence the scale factor of enlargement is 2.

Alternatively, find the scale factor by considering the ratio of the length of a side on the image to the length of the corresponding side on the object, i.e.

$$\frac{A'B'}{AB} = \frac{12}{6} = 2$$

Hence the scale factor of enlargement is 2.

The rectangle $ABCD$ undergoes a transformation to form rectangle $A'B'C'D'$.

a Find the centre of enlargement.
b Calculate the scale factor of enlargement.

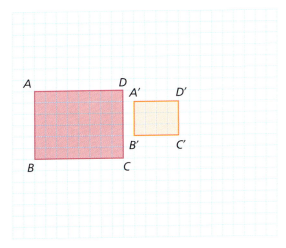

a By joining corresponding points on both the object and the image, the centre of enlargement is found at O.

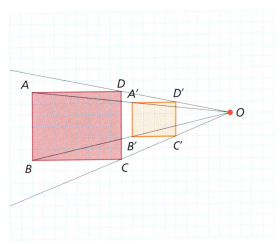

b The scale factor of enlargement $= \dfrac{A'B'}{AB} = \dfrac{3}{6} = \dfrac{1}{2}$.

Note. If the scale factor of enlargement is greater than 1, then the image is larger than the object. If the scale factor lies between 0 and 1, then the resulting image is smaller than the object. In these cases, although the image is smaller than the object, the transformation is still called an enlargement.

Exercise 13.1

Copy each of the following diagrams and:

a   find the centre of enlargement

b   calculate the scale factor of enlargement.

You will need:
- squared paper
- ruler
- coloured pencils

1

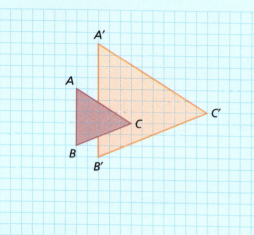

2

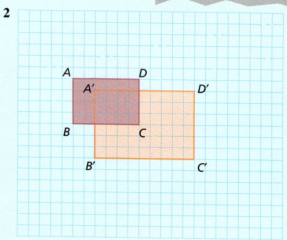

3

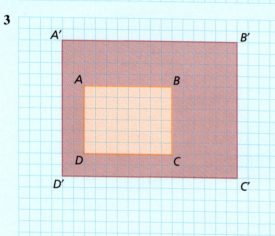

4

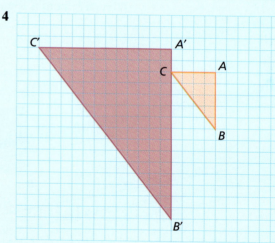

5

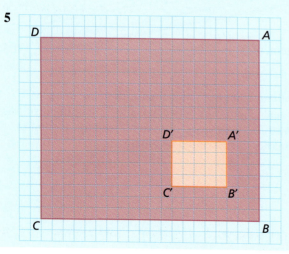

Exercise 13.2

Copy the following diagrams and enlarge the objects by the scale factor given and from the centre of enlargement *O* shown. Grids larger than those shown may be needed.

You will need:
- squared paper
- ruler
- coloured pencils

1

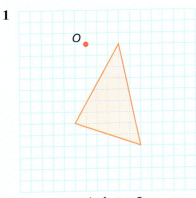

scale factor 2

2

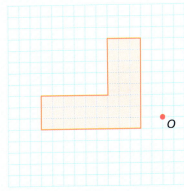

scale factor 2

3

scale factor 3

4

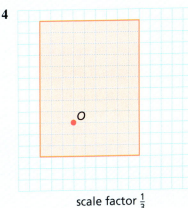

scale factor $\frac{1}{3}$

Translation

If an object is translated, it undergoes a 'straight sliding' movement (or shift). The diagrams below each show object at position *A* undergoing a translation to position *B*.

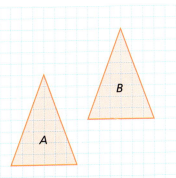

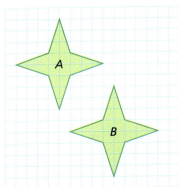

As no rotation is involved, each point on the object moves in the same way to its corresponding point on the image.

You will remember that, to describe the different transformations, certain information needs to be included.

- When describing a reflection, the equation of the line of symmetry is needed.
- When describing a rotation, the centre of rotation, the angle and direction are needed.
- When describing an enlargement, the centre of enlargement and scale factor are needed.

With a translation, the **column vector** is needed.

A translation vector describes how far horizontally and vertically an object is moved. The horizontal and vertical distances are written as a column.

Example

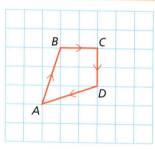

Describe each of the following translations in terms of a column vector.

i) A to B $(\overrightarrow{AB})$ ii) $\overrightarrow{BC}$ iii) $\overrightarrow{CD}$ iv) $\overrightarrow{DA}$

To describe the translation from A to B in terms of a column vector, count the number of units moved horizontally and vertically:

i) $\overrightarrow{AB} = \begin{pmatrix} 1 \\ 3 \end{pmatrix}$ i.e.   1 unit in the x-direction

3 units in the y-direction

ii) $\overrightarrow{BC} = \begin{pmatrix} 2 \\ 0 \end{pmatrix}$

iii) $\overrightarrow{CD} = \begin{pmatrix} 0 \\ -2 \end{pmatrix}$ → Note: the negative here means a vertical movement down

iv) $\overrightarrow{DA} = \begin{pmatrix} -3 \\ -1 \end{pmatrix}$ → Note: the negative here means a horizontal movement left

and a vertical movement down

Vectors can also be defined by a single letter. The direction of the arrow indicates the direction of the translation.

Example Define **a** and **b** using column vectors.

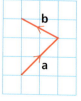

$\mathbf{a} = \begin{pmatrix} 2 \\ 2 \end{pmatrix}$ $\mathbf{b} = \begin{pmatrix} -2 \\ 1 \end{pmatrix}$

Note. When representing vectors by single letters in handwriting you should write them as <u>a</u> or <u>a</u>.

Exercise 13.3

In questions 1 and 2, describe each translation using a column vector.

1

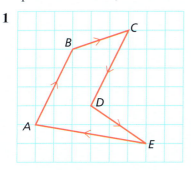

i) $\overrightarrow{AB}$ ii) $\overrightarrow{BC}$

iii) $\overrightarrow{CD}$ iv) $\overrightarrow{DE}$

v) $\overrightarrow{EA}$ vi) $\overrightarrow{AE}$

vii) $\overrightarrow{DA}$ viii) $\overrightarrow{CA}$

ix) $\overrightarrow{DB}$

2

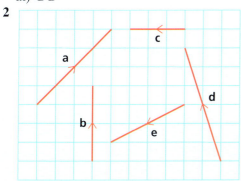

i) **a**   ii) **b**   iii) **c**

iv) **d**   v) **e**   vi) **−b**

vii) **−c**   viii) **−d**   ix) **−a**

3 Draw and label the following vectors **a**, **b**, **c**, etc.

i) $\mathbf{a} = \begin{pmatrix} 2 \\ 4 \end{pmatrix}$ ii) $\mathbf{b} = \begin{pmatrix} -3 \\ 6 \end{pmatrix}$ iii) $\mathbf{c} = \begin{pmatrix} 3 \\ -5 \end{pmatrix}$

iv) $\mathbf{d} = \begin{pmatrix} -4 \\ -3 \end{pmatrix}$ v) $\mathbf{e} = \begin{pmatrix} 0 \\ -6 \end{pmatrix}$ vi) $\mathbf{f} = \begin{pmatrix} -5 \\ 0 \end{pmatrix}$

vii) **−c**   viii) **−b**   ix) **−f**

Examples Give the column vector for each of the translations shown below. Each time the object moves from *A* to *B*.

i) 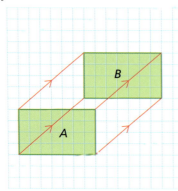 ii)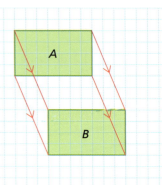

iii) 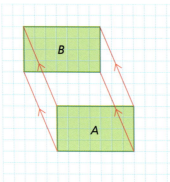 iv)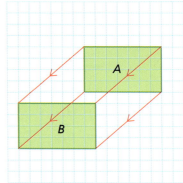

To make translations clearer, lines are drawn from the object to the image. These lines join corresponding points. An arrowhead on the line highlights the direction of the translation.

i) $\begin{pmatrix} 6 \\ 5 \end{pmatrix}$ ii) $\begin{pmatrix} 3 \\ -7 \end{pmatrix}$ iii) $\begin{pmatrix} -3 \\ 7 \end{pmatrix}$ iv) $\begin{pmatrix} -6 \\ -5 \end{pmatrix}$

Exercise 13.4

In each of the following diagrams, object *A* has been translated to i) *B* and ii) *C*. Give the column vector in each case.

1 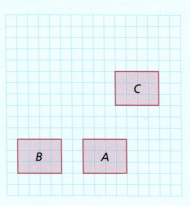   2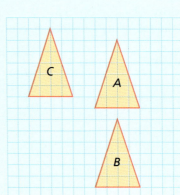

3

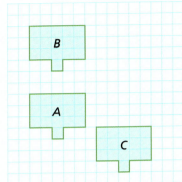

4

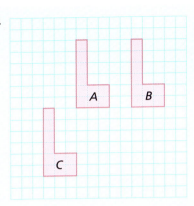

Exercise 13.5

In the following questions, copy the diagram and draw the object. Translate the object by the vector indicated in each case and draw the object in its new position. (Note that a bigger grid than the one shown may be needed.)

You will need:
- squared paper
- ruler
- coloured pencils

1

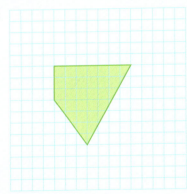

$$\text{Vector} = \begin{pmatrix} 3 \\ 5 \end{pmatrix}$$

2

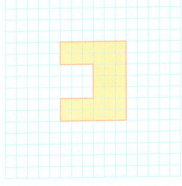

$$\text{Vector} = \begin{pmatrix} 5 \\ -4 \end{pmatrix}$$

3

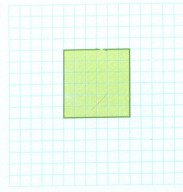

$$\text{Vector} = \begin{pmatrix} -4 \\ 6 \end{pmatrix}$$

4

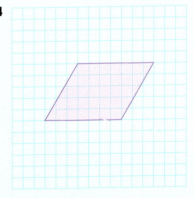

$$\text{Vector} = \begin{pmatrix} -2 \\ -5 \end{pmatrix}$$

5

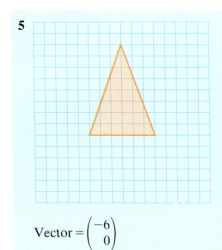

$$\text{Vector} = \begin{pmatrix} -6 \\ 0 \end{pmatrix}$$

6

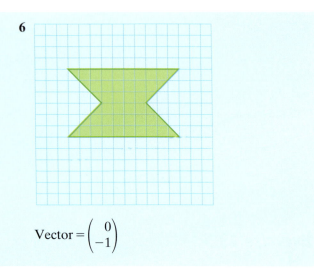

$$\text{Vector} = \begin{pmatrix} 0 \\ -1 \end{pmatrix}$$

Combinations of transformations

An object need not be changed by just one type of transformation. It can be moved by a series of transformations.

Example   A triangle ABC with coordinates $(1, 5)$, $(2, 8)$ and $(3, 4)$ **maps** (is transformed) onto $A'B'C'$ after an enlargement of scale factor 3 from the centre of enlargement $(0, 7)$. $A'B'C'$ is then mapped onto $A''B''C''$ by a reflection in the line $x = 1$.
   a Draw and label the image $A'B'C'$.
   b Draw and label the image $A''B''C''$.

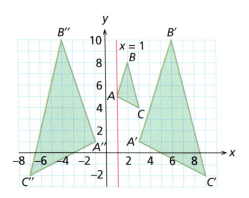

Exercise 13.6

You will need:
• squared paper
• ruler
• coloured pencils

In questions 1 and 2, copy the diagram. For each transformation, draw the images on the same grid and label them clearly. You may need to extend the grid in question 1.

1 a The triangle XYZ is mapped onto $X'Y'Z'$ as a result of a 90° clockwise rotation about the origin.

b Triangle $X'Y'Z'$ is then mapped onto $X''Y''Z''$ by a reflection in the line $y = -x$.

c Finally $X''Y''Z''$ is mapped onto $X'''Y'''Z'''$ by an enlargement of scale factor 2 from a centre of enlargement $(0, 2)$.

Remember:
An **enlargement** changes the size of shapes.
A **translation** moves shapes without changing their shape or size.
A **reflection** moves points to the other side of a mirror line.
A **rotation** turns shapes through a fixed angle.

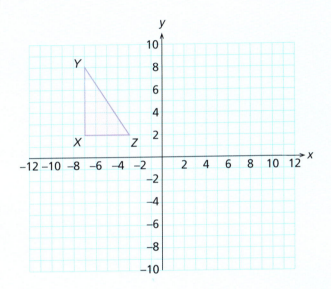

2 The square $ABCD$ is mapped onto $A'B'C'D'$ by a reflection in the line $y = 3$. $A'B'C'D'$ then maps onto $A''B''C''D''$ as a result of a 90° rotation in a clockwise direction about the point $(-2, 5)$.

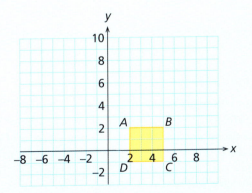

3 In the diagram below, shape *A* is mapped onto shape *B*. Shape *B* is subsequently mapped onto *C*, and *C* onto *D*. Describe fully the transformation which:
a maps *A* onto *B*
b maps *B* onto *C*
c maps *C* onto *D*.

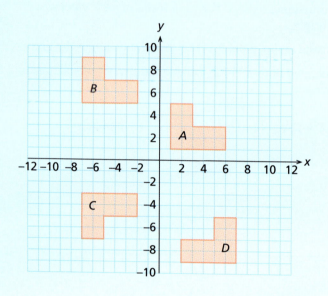

SUMMARY

By the end of this chapter you should know:

- what is meant by an **enlargement** and a **translation**
- that, when describing a **transformation** as an enlargement, two other pieces of information need to be given: the position of the **centre of enlargement** and the **scale factor of enlargement**
- how to find the position of the centre of enlargement given the position of both the object and image
- how to calculate the scale factor of enlargement
- what is meant by a **column vector** and how it is written
- that, when describing a transformation as a translation, the column vector needs to be given
- how to describe different transformations involving a combination of reflection, rotation, enlargement and translation.

Exercise 13A

1 From the diagram below, describe each of the following translations in terms of a column vector.

   a $\overrightarrow{AB}$

   b $\overrightarrow{DA}$

   c $\overrightarrow{CA}$

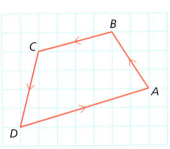

2 Describe each of the following translations in terms of a column vector.

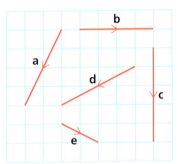

3 Write down the translation vector which maps:

   a rectangle A to rectangle B

   b rectangle B to rectangle C.

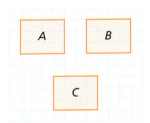

4 Enlarge the triangle below by a scale factor of 2 from centre of enlargement O.

You will need:
• squared paper
• ruler

5 The square $ABCD$ is mapped onto $A'B'C'D'$. $A'B'C'D'$ is then mapped onto $A''B''C''D''$.

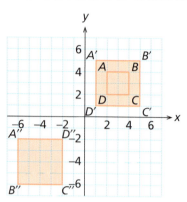

a Describe in full the transformation which maps $ABCD$ onto $A'B'C'D'$.

b Describe in full the transformation which maps $A'B'C'D'$ onto $A''B''C''D''$.

Exercise 13B

1 From the diagram below, describe each of the following translations in terms of a column vector.

a $\overrightarrow{AB}$

b $\overrightarrow{DA}$

c $\overrightarrow{CA}$

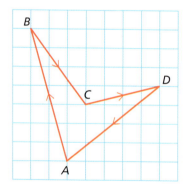

2 Describe each of the translations below in terms of a column vector.

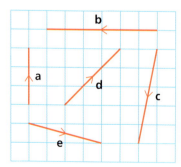

3 Write down the translation which maps:
   a triangle *A* to triangle *B*
   b triangle *B* to triangle *C*.

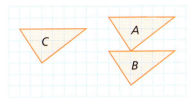

4 Enlarge the rectangle below by a scale factor of 1.5 from centre of enlargement *O*.

You will need:
• squared paper
• ruler

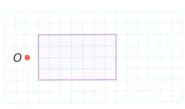

5 The triangle *ABC* is mapped onto *A′B′C′*. *A′B′C′* is then mapped onto *A″B″C″*.

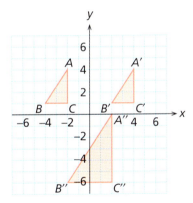

   a Describe in full the transformation which maps *ABC* onto *A′B′C′*.
   b Describe in full the transformation which maps *A′B′C′* onto *A″B″C″*.

Exercise 13C

You will need:
• squared paper
• ruler

The diagram below shows a path drawn as a result of combining four vectors.

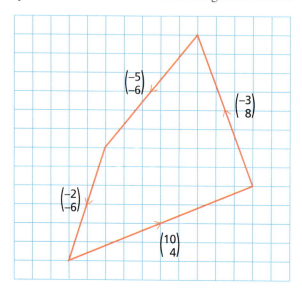

The path follows the order $\begin{pmatrix} 10 \\ 4 \end{pmatrix} \rightarrow \begin{pmatrix} -3 \\ 8 \end{pmatrix} \rightarrow \begin{pmatrix} -5 \\ -6 \end{pmatrix} \rightarrow \begin{pmatrix} -2 \\ -6 \end{pmatrix}$

This order creates a closed path, i.e. it begins and ends in the same place.

1 Rearrange the vectors in a different order. Draw the new path. Is the new path also a closed path?
2 Investigate the different shaped paths that can be drawn by rearranging the four vectors. Are they always closed paths?
3 Explain your findings in question 2.
4 Using your answer to question 3, try to write down a different set of four vectors that will produce a closed path.

Exercise 13D

Using a simple paint package on your computer (such as Microsoft Paint or equivalent), you can create effective tessellating patterns. An example is shown below.

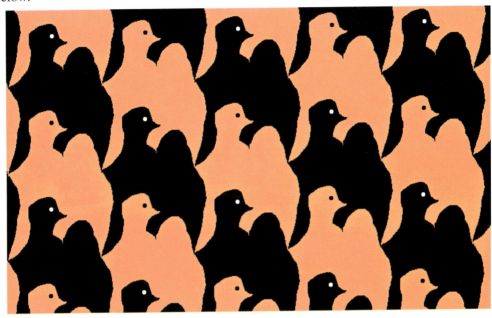

Create your own tessellating pattern.

Exercise 13E

Using the internet or an encyclopaedia as a resource, find some more examples of Escher's work. Try to find out how some of those involving tessellations were drawn.

14 Collecting and analysing data

The microchip affects every aspect of our lives, from the computer sitting on our desktop, to the computers that work in our cars, hospitals and industries.

The most well-known figure of the computing age is Bill Gates, the chairman and chief software architect of Microsoft Corporation. Such is the impact of computing in our everyday lives, that Microsoft had revenues of $22.96 billion for the financial year ending June 2000, and employs more 39 000 people in 60 countries. Bill Gates himself has recently also topped the 'World's richest man' list, with an estimated wealth of around $50 billion.

Computers have become very important in our lives because of their ability to process data very quickly. Although they have to be programmed by humans, they are able to carry out millions of calculations each second – much more quickly than we can!

In *Intermediate 1*, we looked at how data can be analysed using the **mean**, **median** and **mode**. Ways of representing data, using **pie charts**, **bar charts** and **scatter graphs**, were also covered.

In this chapter we will not only look at other ways of representing and analysing data but also look further at how information and computer technology (ICT), in particular spreadsheets, can help. You will probably remember using spreadsheets in many previous exercises.

Note. All spreadsheet screen shots shown in this chapter are taken from Microsoft Excel 2000.

> **Remember:**
> *Mean is the average value, median is the middle value, mode is the most common value.*

> **Remember:**
> *Pie chart*
>
>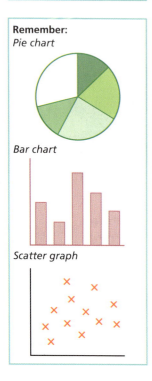
>
> *Bar chart*
>
> *Scatter graph*

Using spreadsheets to plot simple graphs and charts

Spreadsheets speed up the way we carry out calculations and are also capable of producing clear graphs very quickly.

Example   100 pupils were asked to name their favourite chocolate bar. The results are shown below:

chocolate bar	frequency
Mars	16
Kit Kat	24
Twix	31
Snickers	8
Other	21

a   Use a spreadsheet to produce a pie chart of the information.
b   Use a spreadsheet to produce a bar chart of the information.

This example shows what happens if you use Microsoft Excel. If you use other software, your screens may look different.

a • Enter the information into the spreadsheet and select it.

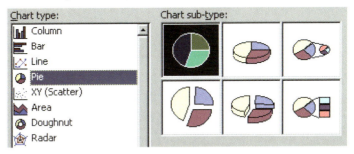

	A	B
1	Mars	16
2	Kit Kat	24
3	Twix	31
4	Snicker	8
5	Other	21

• Select the chart wizard icon from the standard toolbar.

• Select the pie chart option from the 'Chart type' menu.

• Clicking 'Next' will give more information about the data; clicking 'Next' again allows you to give the pie chart a heading.

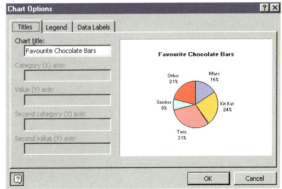

• Clicking on the 'Legend' tab will allow you to decide where you wish to place the key; clicking on the 'Data labels' tab will give you the opportunity to decide how you wish your pie chart to be labelled.
• Clicking on 'Finish' will produce a pie chart similar to the one given below:

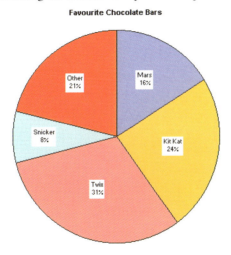

Favourite Chocolate Bars

Note. It is possible to change the final appearance of any part of the chart by simply double clicking on it and then changing it. This was done in the chart above in order to move the labels inside each sector of the pie chart.

b • The first two steps are the same as for drawing a pie chart. However, on the third step, choose either a 'Bar chart' or a 'Column chart' depending on the desired orientation.

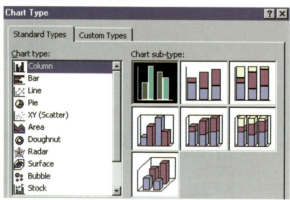

• Clicking on 'Next' twice produces the screen shown below:

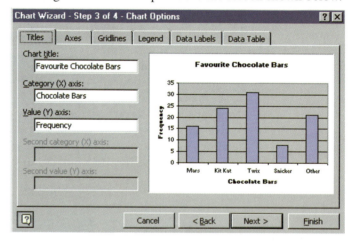

Here the graph can be given a title, and its axes can be given labels. Clicking on each of the other tabs in turn gives access to different attributes of the graph which can be altered. Experiment with each of these to see how they affect the graph.

• The final chart can therefore look like this:

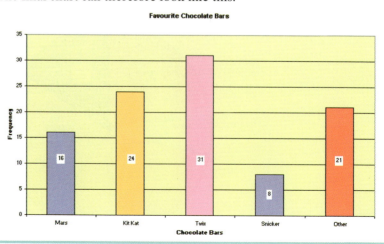

Scatter graphs and correlation

In *Intermediate 1* you learned that there are several types of correlation, depending on the arrangement of the points plotted on the scatter graph.

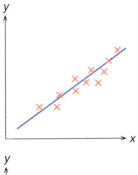

A **strong positive correlation** between the variables x and y. The points lie very close to the line of best fit. As x increases, so does y.

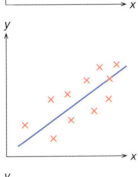

A **weak positive correlation**. Although there is direction to the way the points are lying, they are not tightly packed around the line of best fit. As x increases, y tends to increase too.

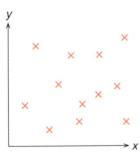

No correlation. As there is no pattern to the way in which the points are lying, there is no correlation between the variables x and y. As a result there can be no line of best fit.

Note. No correlation, or **zero correlation**, does not necessarily imply 'no relationship', but merely 'no linear relationship'.

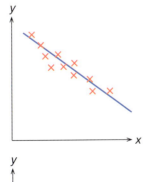

A **strong negative correlation**. The points lie close around the line of best fit. As x increases, y decreases.

A **weak negative correlation**. The points are not tightly packed around the line of best fit. As x increases, y tends to decrease.

Example   A newsagent decides to keep a record of the number of ice creams it sells over a 30 day period, to see if there is any correlation between this and the mean outside temperature. The results are shown in the table below.

temperature (°C)	15	17	18	18	19	20	20	21	22	22	22	23	23	23	24
number of ice creams	8	25	20	30	28	25	34	31	30	32	39	25	32	35	29
temperature (°C)	24	24	24	25	25	25	26	26	26	27	27	27	28	28	30
number of ice creams	34	36	38	28	32	35	37	40	44	33	43	50	35	45	44

a Using a spreadsheet, plot a scatter graph of the data.
b Use the spreadsheet to plot the line of best fit through the points.
c Give the equation of the line of best fit.
d Use the equation of the line of best fit to estimate how many ice creams the newsagent could expect to sell if the temperature were to rise to 35 °C.

a • Enter the data as two columns into the spreadsheet, highlight it and choose the 'chart wizard' icon as shown in the previous worked example above.
 • Choose the 'XY scatter' chart type.

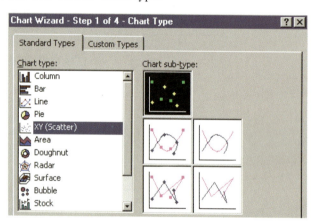

 • The following screens will allow you to label the graph. The finished scatter graph may look similar to the one shown below.

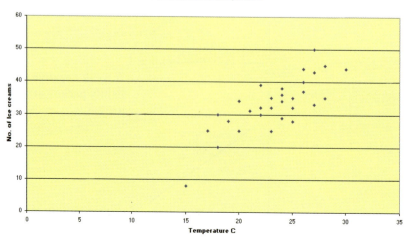

b • You can see from the scatter graph that there appears to be a positive correlation between the two variables, i.e. as the temperature increases, the number of ice creams sold tends also to go up. It is therefore possible to put a line of best fit through the points. To do this, click on Chart followed by Add Trendline ... . This leads to the following screen.

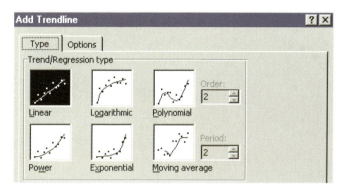

• You can assume that the line of best fit is a straight line, therefore select the 'linear' option. **Note.** To extend the line of best fit either forwards or backwards and to display the equation of the line of best fit on the graph, click on the 'options' tab.
• The final scatter graph, with a line of best fit, and showing the equation, may therefore look similar to the one shown below.

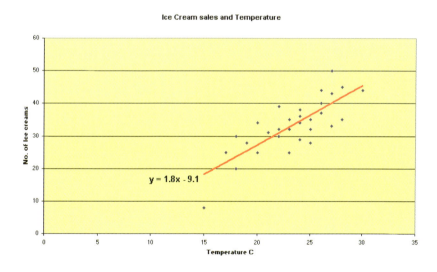

c The equation of the line of best fit is given as $y = 1.8x - 9.1$.
d To estimate the number of ice creams likely to be sold if the temperature were to rise to 35 °C, substitute 35 into the equation for x. This gives:

$$y = 1.8 \times 35 - 9.1$$
$$= 53.9$$

The newsagent can expect to sell approximately 54 ice creams.

Exercise 14.1

You will need:
• computer with spreadsheet package installed
• graph paper

(*Revision*)

1 The table below shows the results of a survey among 72 students to find their favourite sport.

sport	frequency
football	35
tennis	14
volleyball	10
hockey	6
basketball	5
other	2

Use a spreadsheet to illustrate this data on a pie chart.

2 A research project looking at the careers of men and women in Spain produced the following results.

career	male (%)	female (%)
clerical	22	38
professional	16	8
skilled craft	24	16
non-skilled craft	12	24
social	8	10
managerial	18	4

a Compare the two sets of data by using a spreadsheet to illustrate them on a single bar chart.
b Use your chart to make two statements about the data.
c If there are 8 million women in employment in Spain, calculate the number in either professional or managerial employment.

3 A girl completes a survey on the number of advertisements on each page of a 240-page magazine. The results of the survey are given below.

number of advertisements	percentage of pages
0	15
1	35
2	20
3	22
>3	8

a Use a spreadsheet to illustrate this information on a pie chart.
b How many pages had no advertisements?

4 The table below shows the number of passengers (in millions) arriving at four British airports in one year.

airport	number of passengers (millions)
Heathrow	27
Gatwick	16
Stansted	4.5
Luton	2.5

   With the aid of a spreadsheet, illustrate this information in a bar chart.

5 Fifteen pupils take both a maths test and a science test. Their marks are given in the table below.

maths test (%)	67	72	84	41	57	39	92	88	63	57	45	62	68	75	81
science test (%)	71	80	76	52	58	46	88	89	60	50	51	65	74	75	94

   a Use a spreadsheet to plot these results on a scatter graph.
   b Is there a correlation between the two sets of results? If so, use the spreadsheet to draw a linear line of best fit and to give its equation.
   c A pupil was absent for the mathematics exam. If she got 60% in her science exam, use your answers to **b** to estimate her likely score for the maths exam.

So far the **statistical calculations** covered have included:

* the mean – the sum of all the values, divided by the number of values
* the median – the middle value, when all the values have been arranged in numerical order
* the mode – the value that occurs most often.

Moving average

We saw in *Intermediate 1* that the median and the mode as measures of average can sometimes be misleading. The mean can also be unreliable in some circumstances.

The table below shows the daily takings of a stationery counter over a four-week period.

week	day	sales (£)
Week 1	Monday	46
	Tuesday	64
	Wednesday	68
	Thursday	55
	Friday	72
	Saturday	126
	Sunday	103
Week 2	Monday	51
	Tuesday	65
	Wednesday	70
	Thursday	60
	Friday	75
	Saturday	133
	Sunday	110

week	day	sales (£)
Week 3	Monday	54
	Tuesday	65
	Wednesday	72
	Thursday	59
	Friday	78
	Saturday	174
	Sunday	99
Week 4	Monday	57
	Tuesday	60
	Wednesday	75
	Thursday	55
	Friday	80
	Saturday	168
	Sunday	112

This data contains a **cyclical** variation, i.e. takings vary depending on the day of the week. When the data is plotted using a **line graph** its shape looks irregular, as shown below. This may disguise any **trend**.

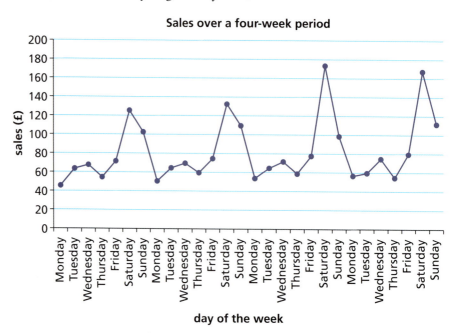

Sales over a four-week period

To identify any trend, a further set of points, known as the **moving average**, is plotted. In this case, because the data follows a seven-day cycle, a seven-point moving average is appropriate. This is shown in the table below.

week	day	sales (£)	moving average
Week 1	Monday	46	
	Tuesday	64	
	Wednesday	68	
	Thursday	55	76.3
	Friday	72	77.0
	Saturday	126	77.1
	Sunday	103	77.4
Week 2	Monday	51	78.1
	Tuesday	65	78.6
	Wednesday	70	79.6
	Thursday	60	80.6
	Friday	75	81.0
	Saturday	133	81.0
	Sunday	110	81.3
Week 3	Monday	54	81.1
	Tuesday	65	81.6
	Wednesday	72	87.4
	Thursday	59	85.9
	Friday	78	86.3
	Saturday	174	85.6
	Sunday	99	86.0
Week 4	Monday	57	85.4
	Tuesday	60	85.7
	Wednesday	75	84.9
	Thursday	55	86.7
	Friday	80	
	Saturday	168	
	Sunday	112	

> You need seven values for a seven-point moving average, so you can't work one out for the last three days.

The first of the moving averages (76.3) is the mean of the first seven days. The second moving average (77.0) is the mean of the Tuesday, Wednesday, Thursday, Friday, Saturday and Sunday of week 1, but also Monday of week 2. The third moving average (77.1) is the mean of the Wednesday, Thursday, Friday, Saturday and Sunday of week 1, but also Monday and Tuesday of week 2.

Note how, in the table, the moving average values are placed alongside the midpoint of the seven numbers they are calculated from. Likewise, when plotted, the moving average values are plotted at the midpoint of the seven values they are calculated from. This is shown in the graph below.

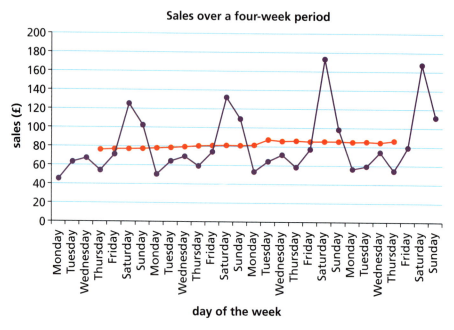

Sales over a four-week period

The moving average line clearly shows that, over the four-week period, average sales rose steadily.

Exercise 14.2

1 'Sox R Us', a sock retail shop, produces a printout of its revenue each quarter. (**Note.** A quarter represents 3 months.) Four years of revenue are shown in the table.

a Plot a line graph, showing quarterly revenue against time.
b Is the data cyclical?
c Calculate a four-point moving average for this data.
d Plot the moving average on your graph.
e Write a brief report stating what the moving average tells you about the revenue figures over these four years.

You will need:
• graph paper

year	quarter	revenue (£)
1998	1	4200
	2	3100
	3	2650
	4	6340
1999	1	3850
	2	3000
	3	2880
	4	6000
2000	1	3700
	2	3400
	3	2550
	4	5880
2001	1	3210
	2	2870
	3	3050
	4	6220

2 The table below shows the percentage of students achieving A*–C in mathematics in a Manchester Secondary school over a 12-year period.

year	1991	1992	1993	1994	1995	1996	1997	1998	1999	2000	2001	2002
percentage	47	55	50	63	52	55	53	58	60	67	62	63

a Plot a line graph of the percentage of students achieving A*–C against time.
b Give a reason why the results show peaks in 1994 and 2000.
c Calculate a five-point moving average for this data.
d Plot the moving average on your graph.
e What does the moving average say about the performance of students in mathematics at that school over the 12 years?
f By drawing a line of best fit through the moving average points and extending it to the year 2003, estimate the moving average value for 2003.
g Using your answer to part **f** above, and showing your working clearly, estimate the percentage achieving A*–C for 2003.

Continuous data

The calculations in *Intermediate 1*, and those so far in this chapter, apply to **discrete data**. Discrete data can only take specific values, for example the number of students in a class. This can only be a whole number, as it would not be possible to have 29.7 students in a class. Shoe sizes also are an example of discrete data as they can only take values such as 4, $4\frac{1}{2}$, 5, $5\frac{1}{2}$, etc. It is not possible to have a shoe size of $4\frac{7}{8}$.

However, some data is known as **continuous data**. This in theory can take any number of values, and is usually associated with data involving measurement, for example time, mass, velocity, length, etc. An example of this would be the winner of a 100 m race being timed at 12.4 seconds. This time is only accurate to 1 d.p. In theory the accuracy of the time could be increased infinitely, i.e. 12.43 seconds or 12.434 seconds or 12.4341 seconds, etc.

Examples   The results below are the times given (in hours:minutes:seconds) for the first 50 people completing the 2001 London marathon.

2:07:11	2:08:15	2:09:36	2:09:45	2:10:45
2:10:46	2:11:42	2:11:57	2:12:02	2:12:11
2:13:12	2:13:26	2:14:26	2:15:34	2:15:43
2:16:25	2:16:27	2:17:09	2:18:29	2:19:26
2:19:27	2:19:31	2:20:00	2:20:23	2:20:29
2:21:47	2:21:52	2:22:32	2:22:48	2:23:08
2:23:17	2:23:28	2:23:46	2:23:48	2:23:57
2:24:04	2:24:12	2:24:15	2:24:24	2:24:29
2:24:45	2:25:18	2:25:34	2:25:56	2:26:10
2:26:22	2:26:51	2:27:14	2:27:23	2:27:37

a Arrange the data into a **grouped frequency** table.
b Using a spreadsheet, plot a grouped frequency graph of the data.
c Calculate the mean time from the grouped frequency table.

a

group	frequency
$2{:}05{:}00 \leqslant t < 2{:}10{:}00$	4
$2{:}10{:}00 \leqslant t < 2{:}15{:}00$	9
$2{:}15{:}00 \leqslant t < 2{:}20{:}00$	9
$2{:}20{:}00 \leqslant t < 2{:}25{:}00$	19
$2{:}25{:}00 \leqslant t < 2{:}30{:}00$	9

Note that, as with discrete data, the groups do not overlap and have equal class intervals. Because the data is continuous, the groups are written using inequalities. The first group includes all times from 2 hours 5 minutes *up to but not including* 2 hours 10 minutes.

> Be careful: make sure your software recognises time format as hours, minutes, seconds, not as decimal.

b A grouped frequency graph constructed from continuous data is different from one constructed from discrete data, in that there are no gaps between the columns. This can be achieved through the spreadsheet as follows.

- Once the graph has been completed, as shown earlier in the chapter, double click on the *x*-axis. This produces the following screen:

- By selecting the 'option' tab, alter the gap width to 0.

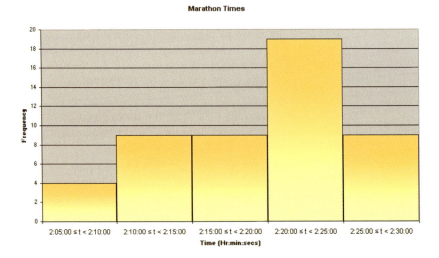

Remember:
Taking the mid-interval values leads to an **estimated** *mean.*

c As with discrete data, calculate the mean time by assuming that each of the values in the group takes the **mid-interval value**.

group	mid-interval value (x)	frequency (f)	fx
2:05:00 $\leqslant t <$ 2:10:00	2:07:30	4	8:30:00
2:10:00 $\leqslant t <$ 2:15:00	2:12:30	9	19:52:30
2:15:00 $\leqslant t <$ 2:20:00	2:17:30	9	20:37:30
2:20:00 $\leqslant t <$ 2:25:00	2:22:30	19	45:07:30
2:25:00 $\leqslant t <$ 2:30:00	2:27:30	9	22:07:30

$$\text{mean} = \frac{8:30:00 + 19:52:30 + 20:37:30 + 45:07:30 + 22:07:30}{50}$$

$$= \frac{116:15:00}{50}$$

$$= 2:19:30$$

The estimated mean time for the first 50 runners is 2 hours 19 minutes 30 seconds.

Exercise 14.3

You will need:
- computer with spreadsheet package installed
- graph paper

1 The table below shows the time taken, in minutes, by 40 pupils to travel to school.

time (min)	frequency
$0 \leqslant t < 5$	6
$5 \leqslant t < 10$	3
$10 \leqslant t < 15$	13
$15 \leqslant t < 20$	7
$20 \leqslant t < 25$	3
$25 \leqslant t < 30$	4
$30 \leqslant t < 35$	4

a Estimate the mean time taken for the pupils to travel to school.
b What is the modal time interval taken for the pupils to travel to school?
c Using a spreadsheet, or otherwise, draw a grouped frequency diagram of the data.

2 On Sundays, Frances helps her father feed their chickens. Over a period of one year she kept a record of how long it took. Her results are shown in the table below.

time (min)	frequency
$0 \leqslant t < 10$	8
$10 \leqslant t < 20$	5
$20 \leqslant t < 30$	8
$30 \leqslant t < 40$	9
$40 \leqslant t < 50$	10
$50 \leqslant t < 60$	12

a Estimate the mean time taken for Frances to feed the chickens.
b What is the modal time interval taken for the chickens to be fed?
c Using a spreadsheet, or otherwise, draw a grouped frequency diagram of the data.

3 The table below shows the ages of 120 people, chosen at random, who take the 6 a.m. train into a city.

age (years)	$0 \leqslant a < 15$	$15 \leqslant a < 30$	$30 \leqslant a < 45$	$45 \leqslant a < 60$	$60 \leqslant a < 75$	$75 \leqslant a < 90$
frequency	3	25	50	32	8	2

The following table shows the ages of 120 people, chosen at random, who take the 11 a.m. train into the city.

age (years)	$0 \leqslant a < 15$	$15 \leqslant a < 30$	$30 \leqslant a < 45$	$45 \leqslant a < 60$	$60 \leqslant a < 75$	$75 \leqslant a < 90$
frequency	21	15	7	5	60	12

a Estimate the mean age of the passengers on each of the two trains.
b Using a spreadsheet, or otherwise, on the same diagram, draw grouped frequency graphs of the two sets of data.
c Compare the two sets of data and give two possible reasons for the differences.

Measures of spread

Simply looking at the mean of a set of data can be misleading. Take a look at the job adverts below for two small shops:

Daily Cooperative

Mean employee salary £15 000

DAILY BREAD

Mean employee salary £15 000

Both appear to be offering similar salaries. However, the individual salaries of the five existing employees in each shop are given below.

Daily Cooperative	£15 000	£15 000	£15 000	£15 000	£15 000
Daily Bread	£10 000	£10 000	£10 000	£10 000	£35 000

The one large salary of £35 000 skews the mean employee salary of the Daily Bread.

Therefore an idea of the **range** of the results would also be useful. The range is worked out by simply subtracting the smallest value from the largest value, for example:

Daily Cooperative range is £15 000 − £15 000 = £0
Daily Bread range is £35 000 − £10 000 = £25 000

This shows that within the Daily Bread shop there is a wide range of salaries; in the Daily Cooperative, all salaries are the same.

A more sophisticated measure of spread involves the use of **cumulative frequency**. The cumulative frequency is also useful for calculating the median of large sets of data, whether grouped or continuous data – something which until now we have not been able to do.

The cumulative frequency is found by adding up the frequencies as we go along, i.e. a 'running total'.

Example The duration of two different brands of battery A and B is tested. Fifty batteries of each type are randomly selected and tested under the same conditions. The duration of each battery is then recorded. The results of the tests are shown in the tables below:

type A		
duration (h)	frequency	cumulative frequency
$0 \leqslant t < 5$	3	3
$5 \leqslant t < 10$	5	8
$10 \leqslant t < 15$	8	16
$15 \leqslant t < 20$	10	26
$20 \leqslant t < 25$	12	38
$25 \leqslant t < 30$	7	45
$30 \leqslant t < 35$	5	50

Remember:
Cumulative frequency is a running total.

type B		
duration (h)	frequency	cumulative frequency
$0 \leqslant t < 5$	1	1
$5 \leqslant t < 10$	1	2
$10 \leqslant t < 15$	10	12
$15 \leqslant t < 20$	23	35
$20 \leqslant t < 25$	9	44
$25 \leqslant t < 30$	4	48
$30 \leqslant t < 35$	2	50

a Plot a cumulative frequency curve for each type of battery.
b Calculate the median duration for each type of battery.

a

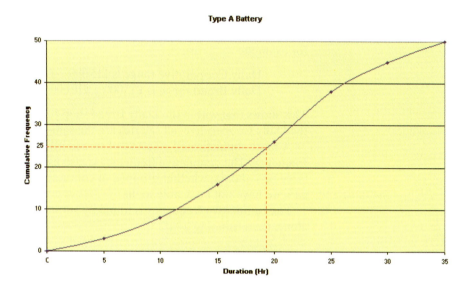

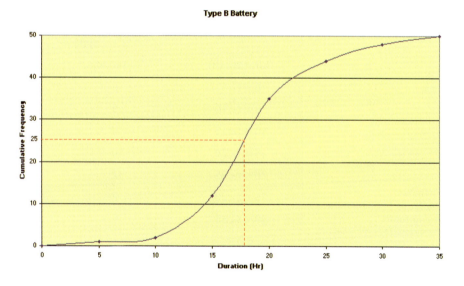

Each value of cumulative frequency is plotted against the **upper boundary** of each class interval and *not* in the middle. When using a spreadsheet, the graph type to choose is 'XY scatter' followed by the 'smooth curve' option.

b The median value is the middle value, i.e. the value which occurs half-way up the cumulative frequency axis. Read these from the graph:

The median for type A batteries ≈ 19 hours.

The median for type B batteries ≈ 18 hours.

Exercise 14.4

You will need:
- computer with spreadsheet package installed
- graph paper

1 Sixty athletes enter a cross-country race. Their finishing times are recorded and are shown in the table below.

finishing time (h)	$0.5 \leqslant t < 1$	$1 \leqslant t < 1.5$	$1.5 \leqslant t < 2$	$2 \leqslant t < 2.5$	$2.5 \leqslant t < 3$	$3 \leqslant t < 3.5$
frequency	0	6	34	16	3	1
cumulative frequency						

   a Copy the table and calculate the values for the cumulative frequency.
   b Using a spreadsheet or otherwise, draw a cumulative frequency curve of the results.
   c Using your graph, estimate the median finishing time.

2 Three mathematics classes take the same test in preparation for their final exam. Their raw scores are shown below.

Class A	12	21	24	30	33	36	42	45	53	53	57	59	61	62	74	88	92	93
Class B	48	53	54	59	61	62	67	78	85	96	98	99						
Class C	10	22	36	42	44	68	72	74	75	83	86	89	93	96	97	99	99	

   a Using class intervals of $0 \leqslant x < 20$, $20 \leqslant x < 40$, etc., draw up a grouped frequency and cumulative frequency table for each class.
   b Using a spreadsheet or otherwise, draw a cumulative frequency curve for each class.
   c Using your graphs, estimate the median score for each class.

Keep your results for questions 2 and 3. You will need them again in exercise 14.5

3 The table below shows the height of students in a class over a three-year period.

height (cm)	frequency 2000	frequency 2001	frequency 2002
$150 \leqslant h < 155$	6	2	2
$155 \leqslant h < 160$	8	9	6
$160 \leqslant h < 165$	11	10	9
$165 \leqslant h < 170$	4	4	8
$170 \leqslant h < 175$	1	3	2
$175 \leqslant h < 180$	0	2	2
$180 \leqslant h < 185$	0	0	1

   a Construct a cumulative frequency table for each year.
   b Using a spreadsheet or otherwise, draw the cumulative frequency curve for each year.
   c Using your graphs, estimate the median height for each year.

Quartiles and the interquartile range

The cumulative frequency axis can also be represented in terms of **percentiles**. A percentile scale divides the cumulative frequency scale into hundredths. The maximum value of cumulative frequency is known as the 100th percentile. Similarly the median, being the middle value, is called the 50th percentile (Q_2). The 25th percentile is known as the **lower quartile** (Q_1) and the 75th percentile is called the **upper quartile** (Q_3).

The range of a distribution, as has already been mentioned, is found by subtracting the lowest value from the highest value. Sometimes this will give a useful result, but often it will not. A better measure of spread is given by looking at the spread of the middle half of the results, i.e. the difference between the upper and lower quartiles. This result is known as the **interquartile range**.

The graph below displays the terms mentioned above.

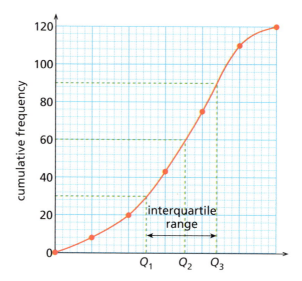

Key
Q_1 = lower quartile
Q_2 = median
Q_3 = upper quartile
$Q_3 - Q_1$ = interquartile range

Example   Consider again the two types of batteries A and B discussed earlier.

a Using the graphs, estimate the upper and lower quartiles for each battery.
b Calculate the interquartile range for each battery.
c Based on these results, how might the manufacturers advertise the two types of batteries?

a

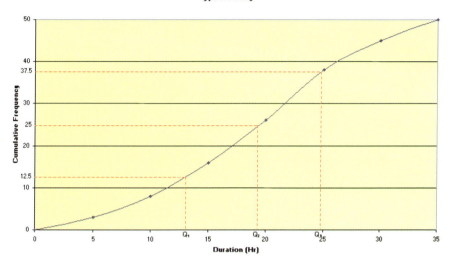

Type A Battery

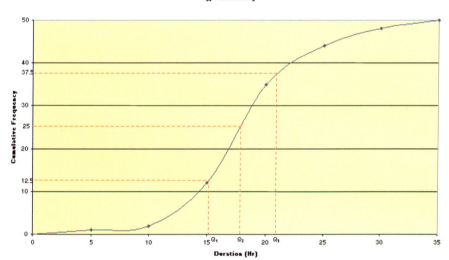

Type B Battery

Lower quartile for type A ≈ 13 h.   Lower quartile for type B ≈ 15 h.
Upper quartile for type A ≈ 25 h.   Upper quartile for type B ≈ 21 h.

b Interquartile range for type A ≈ 12 h.
Interquartile range for type B ≈ 6 h.

c Type A:
On average, this is the longer-lasting battery (as the median for type A is greater than for type B)
Type B:
This is the more reliable battery (as the interquartile range is less).

Exercise 14.5

1 Using the results obtained from question 2 in exercise 14.4,
  a find the interquartile range of each of the classes taking the mathematics test
  b analyse your results and write a brief summary comparing the three classes.
2 Using the results obtained from question 3 in the exercise 14.4,
  a find the interquartile range of the students' heights each year
  b analyse your results and write a brief summary comparing the three years.
3 Forty boys enter for a school javelin competition. The distances thrown are recorded below:

distance thrown (m)	$0 \leqslant d < 10$	$10 \leqslant d < 20$	$20 \leqslant d < 30$	$30 \leqslant d < 40$	$40 \leqslant d < 50$
frequency	4	9	15	10	2

  a Construct a cumulative frequency table for the above results.
  b Using a spreadsheet or otherwise, draw a cumulative frequency curve.
  c If the top 25% of boys are considered for the final, use the graph to estimate the qualifying distance.
  d Calculate the interquartile range of the throws.
  e Calculate the median distance thrown.
4 The mass of two different types of oranges are compared. Eighty oranges are randomly selected from each type and weighed. The results are shown below.

type A		type B	
mass (g)	frequency	mass (g)	frequency
$75 \leqslant x < 100$	4	$75 \leqslant x < 100$	0
$100 \leqslant x < 125$	7	$100 \leqslant x < 125$	16
$125 \leqslant x < 150$	15	$125 \leqslant x < 150$	43
$150 \leqslant x < 175$	32	$150 \leqslant x < 175$	10
$175 \leqslant x < 200$	14	$175 \leqslant x < 200$	7
$200 \leqslant x < 225$	6	$200 \leqslant x < 225$	4
$225 \leqslant x < 250$	2	$225 \leqslant x < 250$	0

  a Construct a cumulative frequency table for each type of orange.
  b Using a spreadsheet or otherwise, draw a cumulative frequency graph for each type of orange.
  c Calculate the median mass for each type of orange.
  d Using your graphs, estimate:
    i) the lower quartile
    ii) the upper quartile
    iii) the interquartile range for each type of orange.
  e Write a brief report comparing the two types of orange.

5 Two competing brands of batteries are compared. 100 batteries of each brand are tested and the duration of each is recorded. The results of the tests are shown in the cumulative frequency graphs below.

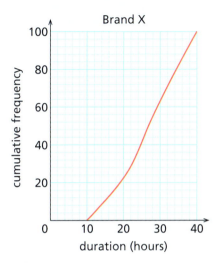

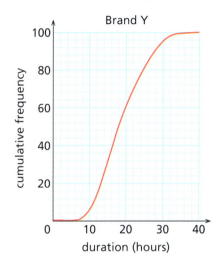

a The manufacturers of brand X claim that on average their batteries will last at least 40% longer than those of brand Y. Showing your method clearly, decide whether this claim is true.
b The manufacturers of brand X also claim that their batteries are more reliable than those of brand Y. Is this claim true? Show your working clearly.

Box plots

So far we have seen how cumulative frequency curves enable us to look at how data is **dispersed** (spread out) by working out the upper and lower quartiles and also the interquartile range.

Box plots provide another visual way of representing the spread of data. The diagram below demonstrates what a 'typical' box plot looks like and also highlights its main features:

> Box plots are sometimes called box-and-whisker diagrams.

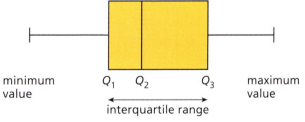

The box plot shows the five main features of the data, i.e. the minimum and maximum values, the upper and lower quartiles, and the median (sometimes known as the five **indicators**). The box itself represents the middle 50% of the data (the interquartile range).

Example The shoe sizes of 15 boys and 15 girls from the same class are recorded in the frequency table below.

shoe size	5	$5\frac{1}{2}$	6	$6\frac{1}{2}$	7	$7\frac{1}{2}$	8	$8\frac{1}{2}$	9	$9\frac{1}{2}$
frequency (boys)	0	0	1	2	1	2	3	4	1	1
frequency (girls)	1	3	4	4	1	1	1	0	0	0

> **Remember:**
> *The median is the middle value once the data is ordered.*

a Calculate the lower quartile, median and upper quartile shoe sizes for the boys and girls in the class.

b Compare this data using two box plots (one for boys and one for girls).

c What conclusions can be made from the box plots?

a In *Intermediate 1* we showed that for discrete data the median position is given by the formula $\dfrac{n+1}{2}$ where n represents the number of values.

Similarly the position of the lower quartile can be calculated using the formula $\dfrac{n+1}{4}$ and the upper quartile by the formula $\dfrac{3(n+1)}{4}$.

$$\text{lower quartile boy} = \frac{15+1}{4} = 4\text{th} \qquad \text{i.e. } Q_1 = 7$$

$$\text{median boy} = \frac{15+1}{2} = 8\text{th} \qquad \text{i.e. } Q_2 = 8$$

$$\text{upper quartile boy} = \frac{3(15+1)}{4} = 12\text{th} \quad \text{i.e. } Q_3 = 8\tfrac{1}{2}$$

$$\text{lower quartile girl is the 4th} \qquad \text{i.e. } Q_1 = 5\tfrac{1}{2}$$

$$\text{median girl is the 8th} \qquad \text{i.e. } Q_2 = 6$$

$$\text{upper quartile girl is the 12th} \qquad \text{i.e. } Q_3 = 6\tfrac{1}{2}$$

b For box plots it is also necessary to know the maximum and minimum values.

min. boy shoe size is 6
max. boy shoe size is $9\frac{1}{2}$
min. girl shoe size is 5
max. girl shoe size is 8

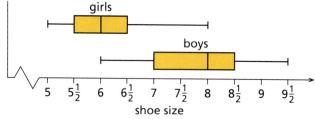

Note. There is no scale on the y-axis as it is not relevant in a box plot. The height of a box plot does not have a meaning.

c The overall range of data is greater for boys than it is for girls.
The interquartile range for girls is less than for boys, i.e. the middle 50% of girls have a narrower range of shoe size than the middle 50% of boys.

Using a graphical calculator

Some graphical calculators are able to calculate Q_1, Q_2 and Q_3 and are also able to draw box plots. The Texas Instruments TI-83 is one such calculator.

The screens below demonstrate how the previous worked example can be worked out using a graphical calculator.

> *Other calculators may work differently. Be careful!*

- Enter the data in the frequency table as three lists.

  The list L_1 contains the shoe sizes, while L_2 and L_3 contain the frequency of boys and girls respectively with each shoe size.

L1	L2	L3
5	0	1
5.5	0	3
6	1	4
6.5	2	4
7	1	1
7.5	2	1
8	3	1
8.5	4	0
9	1	0
9.5	1	0

L1(1)=5

- Once the data has been entered it is possible to analyse the boys' and girls' information in turn. Press the STAT key, select 'CALC' then select the type of data, in this case a one-variable statistic, for the boys' results for example.

```
EDIT CALC TESTS
 1:1-Var Stats
 2:2-Var Stats
 3:Med-Med
 4:Lin Reg(ax+b)
 5:QuadReg
 6:CubicReg
 7↓QuartReg
```

- The next screen asks you to identify which list the data is stored in. In the case of the boys, the shoe sizes are in L_1 and the frequency in L_2.

```
1-Var Stats L1,L2
```

Press the ENTER key to obtain a screen full of information.

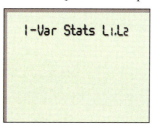

```
1-Var Stats
x̄=7.833333333
Σx=117.5
Σx²=934.25
Sx=.9940297974
σx=.9603240194
n=15
minX=6
Q1=7
Med=8
Q3=8.5
maxX=9.5
```

$\bar{x}$ represents the mean shoe size for boys

n represents the number of boys
minx is the smallest boys' shoe size
Q_1 represents the lower quartile
Med is the median shoe size for boys
Q_3 represents the upper quartile
maxX is the largest boys' shoe size

- Repeat similar steps to the ones shown above for an analysis of the girls' data.
- Once the data has been entered as lists, you can draw box plots. To do this press the STAT PLOT key. This leads to the following screen:

This will enable you to plot both box plots on the same screen.

- By pressing ENTER you are then able to select the type of graph for each of the plots.

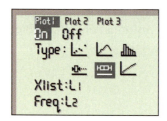

The options on this screen show that **Plot1** is **On**. Its type is a **box plot**. The shoe sizes are in list L_1 and the boys' frequency is in L_2.

A similar process can be followed to produce a box plot of the girls' data in **Plot2**.

- Once the previous steps have been followed, press the GRAPH key to produce the following screen:

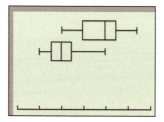

- Using the TRACE key enables you to read important data off from the screen.

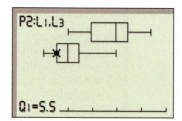

This screen shows the cursor on the bottom box plot. The position of the cursor is given at the bottom of the screen, i.e. $Q_1 = 5.5$.

Exercise 14.6

Try at least one of these questions without a graphical calculator.

Using a graphical calculator or otherwise, answer the following questions.

1 A football team records the number of goals it scored and how many goals it let in, in each of 20 matches. The results are shown in the table below.

number of goals	0	1	2	3	4	5
frequency of goals scored	6	9	3	1	0	1
frequency of goals let in	3	3	8	3	3	0

   a For goals scored and goals let in, work out:
    i) the mean
    ii) the median
    iii) the lower quartile
    iv) the upper quartile
    v) the interquartile range.
   b On the same diagram, represent the data for goals scored and goals let in as box plots.
   c Write a brief report about what the box plots tell you about the team's results.

2 Two competing holiday resorts record the number of hours of sunshine they have each day during the month of August. The results are shown below.

hours of sunshine	4	5	6	7	8	9	10	11	12
resort A	1	2	3	5	5	4	4	4	3
resort B	0	0	0	4	12	10	5	0	0

   a For each resort, work out the number of hours of sunshine represented by:
    i) the mean
    ii) the median
    iii) the lower quartile
    iv) the upper quartile
    v) the interquartile range.
   b On the same diagram, represent the data for both resorts using box plots.
   c Based on the data and referring to your box plots, explain which resort you would choose to go to for a beach holiday in August.

3 A teacher decides to tackle the problem of students arriving late to his class. To do this, he starts by timing how late they are. He records the results to the nearest minute. These are shown in the table below.

number of minutes late	0	1	2	3	4	5	6	7	8	9	10
number of students	6	4	4	5	7	3	1	0	0	0	0

After two weeks of trying to improve the situation, the teacher records a new set of results. These are shown below.

number of minutes late	0	1	2	3	4	5	6	7	8	9	10
number of students	14	7	4	1	1	1	0	0	1	0	1

The teacher decides to analyse these sets of data using box plots. By carrying out any necessary calculations and drawing the relevant box plots, decide whether his strategy has improved student punctuality. Give detailed reasons for your answer.

Stem-and-leaf diagrams

Stem-and-leaf diagrams are a special sort of bar chart, in which the bars are made from the data itself. This has the advantage that the original data can be recovered easily from the diagram. (This is not possible with a box plot, cumulative frequency curve, or a bar chart.)

The **stem** is the first digit of the numbers, so if the numbers are 63, 65, 67, 68, 69 then the stem is 6.

The **leaves** are the remaining digits in ascending order. In the example above, the leaves would be 3, 5, 7, 8 and 9.

Example The ages of the people on a coach transferring them from an airport to a ski resort are as follows:

22	24	25	31	33	23	24	26	37	42
40	36	33	24	25	18	20	27	25	33
28	33	35	39	40	48	27	25	24	29

Display the data on a stem-and-leaf diagram.

This data can be shown on a stem-and-leaf diagram as shown below.

```
1 | 8                                            key
2 | 0 2 3 4 4 4 4 5 5 5 5 6 7 7 8 9         2 | 5 means 25
3 | 1 3 3 3 3 5 6 7 9
4 | 0 0 2 8
```

Note that, on the right, the key states what the stem means.
If the data were 1.8, 2.7, 3.2, etc., the key would state that **2** | 7 means 2.7.
Similarly if the data were 120, 230, 360, etc., the key would state that **2** | 3 means 230.

Exercise 14.7

1 A test in mathematics is marked out of 40. The scores for a class of 32 students are shown below. Display the data on a stem-and-leaf diagram.

24	27	30	33	26	27	28	39
21	18	16	33	22	38	33	21
16	11	14	23	37	36	38	22
28	15	9	17	28	33	36	34

2 A basketball team played 24 matches in the 2002 season. Display the scores below on a stem-and-leaf diagram.

62	48	85	74	63	67	71	83
46	52	63	65	72	76	68	58
54	46	88	55	46	52	58	54

3 A class of 27 students was asked to draw a line 8 cm long without a ruler. The lines were then measured and the data recorded below. Illustrate this data on a stem-and-leaf diagram.

8.8	6.2	8.3	7.9	8.0	5.9	6.2	10.0	9.7
7.9	5.4	6.8	7.3	7.7	8.9	10.4	5.9	8.3
6.1	7.2	8.3	9.4	6.5	5.8	8.8	8.0	7.3

Back-to-back diagrams

Stem-and-leaf diagrams are often used as an easy way to compare two sets of data. The leaves are usually put 'back-to-back' on either side of the stems.

Example In the previous example, the stem-and-leaf diagram for the data for ages of people on a coach to a ski resort is as shown below. The data is easily accessible.

```
1 | 8                                        key
2 | 0 2 3 4 4 4 4 5 5 5 5 6 7 7 8 9         2 | 5 means 25
3 | 1 3 3 3 3 5 6 7 9
4 | 0 0 2 8
```

A second coach from the airport is taking people to a golfing holiday. The ages of the people are shown below.

43	46	52	61	65	38	36	28	37	45
69	72	63	55	46	34	35	37	43	48
54	53	47	36	58	63	70	55	63	64

Display the two sets of data on a back-to-back stem-and-leaf diagram.

The two sets of data can be compared on a back-to-back diagram as shown below.

```
        golf                          skiing
                      8 | 1 | 8
                    8 | 2 | 0 2 3 4 4 4 4 5 5 5 5 6 7 7 8 9
      8 7 7 6 6 5 4 | 3 | 1 3 3 3 3 5 6 7 9
      8 7 6 6 5 3 3 | 4 | 0 0 2 8
          8 5 5 4 3 2 | 5 |                      key
        9 5 4 3 3 3 1 | 6 |                       3 | 5 means 35
                  2 0 | 7 |
```

Exercise 14.8

1 Write three sentences commenting on the back-to-back diagram in the example above.

2 The basketball team in question 2 of exercise 14.7 had replaced their team coach at the end of the 2001 season. The scores for the 24 matches of the previous season were:

82	32	88	24	105	63	86	42
35	88	78	106	64	72	88	26
35	41	100	48	54	36	28	33

Display the scores from the two seasons back-to-back and comment on the diagram.

3 The mathematics test results shown in question 1 of exercise 14.7 were for test B. Test A had already been set, marked and the teacher had gone over some of the questions on the board. The marks out of 40 for test A were as follows:

22	18	9	11	38	33	21	14
16	8	12	37	39	25	23	18
34	36	23	16	14	12	22	29
33	35	12	17	22	28	32	39

Draw a back-to-back stem-and-leaf diagram for both test scores and comment on the diagram.

Flow charts

The types of graph we have used so far represent ways of displaying data. **Flow charts** can be used to sort data or to show a sequence of instructions. The flow chart below is an example of how one can be used to categorise pupils in a class by gender and height.

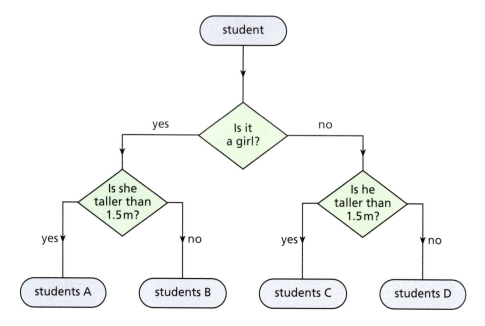

When using a flow chart for classification, two main types of box are used.

These are used for 'start' or 'read' boxes. In the flow chart above, all students will start at the top box and students' names will be read from the bottom ones.

The rhombus shapes are used when asking questions leading to a decision. It can be deduced from the above flow chart that the students in box A will be those girls taller than 1.5 m. The students in box D will be those boys shorter than 1.5 m.

Exercise 14.9

1 Design a flow chart that will organise the students in your class by gender and hair colour.
2 Design a flow chart that will organise the following animals into categories: lion, dog, cobra, dolphin, whale, tiger, domestic cat, sardine, boa constrictor, swordfish and wolf.

3 Below are a number of geometric shapes.

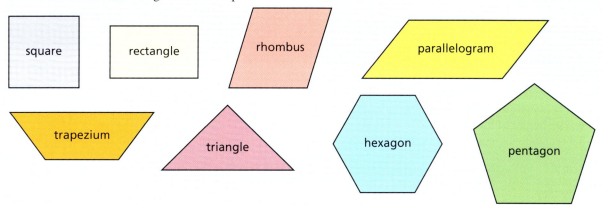

Design a flow chart that will individually identify each of the shapes above.

SUMMARY

By the time you have finished this chapter you should know:

- how to use a spreadsheet to produce **bar charts**, **pie charts** and **scatter graphs**
- that a **moving average** can be used to even out **cyclical** variations in data
- how to calculate and plot a moving average
- whether data is **continuous** or **discrete**; for example the number of students in a class is discrete data, while the height of students in a class is continuous data
- how to calculate the mean of grouped discrete and continuous data
- what is meant by **cumulative frequency** and how it is calculated, i.e. by keeping a running total of the frequencies
- how to plot a cumulative frequency curve, i.e. know that points are plotted at the **upper boundary** of each group
- how to read off the **upper** and **lower quartiles** and the **median** from the cumulative frequency curve, i.e.

  lower quartile (Q_1)
  median (Q_2)
  upper quartile (Q_3)

- what is meant by the **interquartile range** and how it is calculated, i.e. the interquartile range is the spread of the middle 50% of the data, calculated by subtracting the lower quartile from the upper quartile ($Q_3 - Q_1$)

- how to draw a **box plot** and understand what each part of a box plot represents

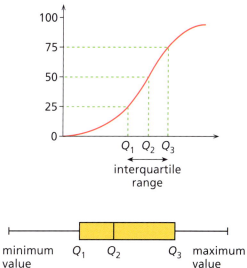

- how to use a graphical calculator, if available, to carry out statistical calculations
- how to draw a **stem-and-leaf diagram**, and how to use one to compare two sets of data
- what is meant by a flow chart and how to create one to classify information.

Exercise 14A

You will need:
• graph paper

1 Identify which of the following types of data are discrete and which are continuous.
  a The number of cars passing the school gate each hour
  b The time taken to travel to school each morning
  c The speed at which students run in a race
  d The wingspan of butterflies
  e The height of buildings

2 Thirty students sit a French exam. Their marks are given as percentages and are shown in the table below.

mark	$20 \leqslant x < 30$	$30 \leqslant x < 40$	$40 \leqslant x < 50$	$50 \leqslant x < 60$	$60 \leqslant x < 70$	$70 \leqslant x < 80$	$80 \leqslant x < 90$	$90 \leqslant x < 100$
frequency	2	3	5	7	6	4	2	1

  a Construct a cumulative frequency table of the above results.
  b Draw a cumulative frequency curve of the results.
  c Using the graph, estimate a value for:
    i) the median
    ii) the upper and lower quartiles
    iii) the interquartile range.

3 A business woman travels to work in her car each morning in one of two ways, either using the country lanes or using the motorway. She records the time taken to travel to work each day. The results are shown in the table below.

time (min)	$10 \leqslant t < 15$	$15 \leqslant t < 20$	$20 \leqslant t < 25$	$25 \leqslant t < 30$	$30 \leqslant t < 35$	$35 \leqslant t < 40$	$40 \leqslant t < 45$
motorway frequency	3	5	7	2	1	1	1
country lanes frequency	0	0	9	10	1	0	0

  a Complete a cumulative frequency table for each of the sets of results shown above.
  b Using your cumulative frequency tables, plot two cumulative frequency curves: one for the time taken to travel to work using the motorway, the other for the time taken to travel to work using country lanes.
  c Use your graphs to work out the following for each method of travel:
    i) the median travelling time
    ii) the upper and lower quartile travelling times
    iii) the interquartile range for the travelling times.
  d With reference to your graphs or calculations, explain which is the most reliable way of getting to work.
  e If the business woman had to get to work one morning within 25 minutes of leaving home, which way would you recommend she take? Explain your answer fully.

4 Two different classes take a maths test. One class is a maths set in which the students are of similar ability, the other is a mixed-ability maths class. The results of the tests for each class are presented using box plots.
Two box plots are shown below.

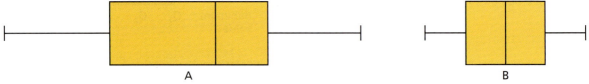

A   B

Explain clearly, giving your reasons, which of the two box plots is *likely* to belong to the mixed-ability maths class and which is likely to belong to the maths set.

5 Twenty students do three long jumps. The best result for each student (in metres) is recorded below.

| 4.3 | 5.4 | 4.3 | 4.0 | 3.8 | 5.1 | 3.6 | 5.5 | 6.2 | 4.7 |
| 5.2 | 3.8 | 2.4 | 4.7 | 3.9 | 5.6 | 5.8 | 4.7 | 3.3 | 2.9 |

The students were then coached in the technique of long jump and given three further jumps. The distance of the best jump was recorded below.

| 4.7 | 5.9 | 4.8 | 4.6 | 4.5 | 5.3 | 5.2 | 5.5 | 6.3 | 4.9 |
| 5.2 | 4.9 | 5.6 | 5.3 | 6.8 | 5.4 | 5.8 | 5.4 | 4.3 | 5.5 |

Draw back-to-back stem-and-leaf diagrams of 'before' and 'after' coaching, and comment on your diagram.

Exercise 14B

You will need:
• graph paper

1 Identify which of the following types of data are discrete and which are continuous.
  a The number of goals scored in a hockey match
  b Dress sizes
  c The time taken to fly from London to Belfast
  d The price of a kilogram of carrots
  e The speed of a police car

2 Four hundred students sit their GCSE Mathematics exam. Their marks (as percentages) are shown in the table.
  a Copy and complete the table by calculating the cumulative frequency.
  b Draw a cumulative frequency curve of the results.
  c Using the graph, estimate a value for:
   i) the median exam mark
   ii) the upper and lower quartiles
   iii) the interquartile range.

mark (%)	frequency	cumulative frequency
31–40	21	
41–50	55	
51–60	125	
61–70	74	
71–80	52	
81–90	45	
91–100	28	

3 Eight hundred students sit an exam. Their marks (as percentages) are shown in the table.
  a Copy and complete the table by calculating the cumulative frequency.
  b Draw a cumulative frequency curve of the results.
  c An A grade is awarded to any student achieving at or above the upper quartile. Using your graph, identify the minimum mark required for an A grade.
  d Any student below the lower quartile is considered to have failed the exam. Using your graph, identify the minimum mark needed so as not to fail the exam.
  e How many students failed the exam?
  f How many students achieved an A grade?

mark (%)	frequency	cumulative frequency
1–10	10	
11–20	30	
21–30	40	
31–40	50	
41–50	70	
51–60	100	
61–70	240	
71–80	160	
81–90	70	
91–100	30	

4 Draw a flow chart that will result in the classification of each of the following angles.

An acute angle
A right angle
An obtuse angle
A reflex angle

Exercise 14C

Below are some strange box plots.

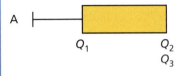

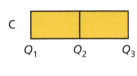

1 Which of the above box plots are possible?
2 For those that are possible, write some data that would produce a box plot of that type.

Exercise 14D

You will need:
• computer with spreadsheet package installed

1 The table below was taken from a UNICEF report in 1996.

percentage of primary school age boys and girls out of school		
region	boys	girls
Sub-Saharan Africa	44	50
Middle East and North Africa	13	22
Asia and Pacific	15	22
Americas	8	7
Europe	8	7

Source: The Progress of Nations 1996, UNICEF

Using spreadsheets, represent this information graphically. Write a brief report about what this information says about the percentages of boys and girls out of (not attending) primary schools.

2 The tables below were produced by the Red Cross in 1997.

The first table shows the number and type of natural disaster that occurred in the world during the years 1970–1994. The second table shows what effect these disasters had on the populations of those regions.

number of natural disasters by region and type over 25 years (1970–1994)						
	Africa	Americas	Asia	Europe	Oceania	total
earthquake	41	135	252	165	85	678
drought and famine	296	53	88	16	16	469
flood	184	382	653	154	135	1508
landslide	12	90	99	21	10	232
high wind	84	454	685	228	199	1650
volcano	9	33	46	16	6	110
other	205	90	189	94	6	593
total	831	1237	2012	694	457	5240

number of people affected by natural disasters (annual average over 25 years 1970–1994)						
	Africa	Americas	Asia	Europe	Oceania	total
killed	76 485	8 988	55 922	2 240	94	143 729
injured	1 017	15 180	37 288	3 475	135	57 095
affected	11 450 827	4 481 691	111 473 882	561 580	653 580	128 621 560
homeless	256 871	308 359	4 334 807	64 965	14 077	4 979 079
total	11 785 200	4 814 218	115 901 899	632 260	667 886	133 801 463

Source: World Disasters Report 1997, International Federation of Red Cross and Red Crescent Societies

Prepare a report for the Red Cross, which describes some of the findings shown in the tables above. Ensure that your report makes comparisons between the data in both tables. Using spreadsheets, include any relevant graphs in your report.

Exercise 14E

You will need:
- computer with internet access

Florence Nightingale became famous as the 'lady with the lamp' because of her work as a nurse in the Crimean War. However, she also made a major contribution to public health through her analysis of statistics.

Using the internet as a resource, see if you can find further information on her statistical work and the impact that it had.

Probability

Hoping for a winning streak...Linda Barrett and her surprise Lottery ticket

Linda game for a laugh

By CHRISTEN PEARS

A WOMAN thought she had hit a winning streak when she bought a lucky dip ticket for the National Lottery.

Linda Barrett, of Field End, Balsham, bought the ticket for the draw on February 14 but only took it out of her purse to check the numbers at the weekend.

She was amazed to see the machine had chosen 24, 25, 26, 27, 28 and 29.

She said: "I know the odds of winning the first Lottery prize are incredibly high, but the odds of having a random selection of numbers in sequence must be equally amazing.

"I had to look twice at the ticket and I thought my luck must be in and it would be the start of a winning streak.

"Unfortunately, it's not the case but it has given us all a bit of fun.

"It's not often that I buy a ticket but I'll certainly try again – you never know."

Mrs Barrett's numbers certainly look unusual but are they any less likely than any other combination of six numbers? A reply to this was also printed in the newspaper.

Mrs Barrett's numbers seem remarkable, but the chance of obtaining her particular consecutive sequence is the same for any other collection of numbers – one in 13,983,816.

However, because there are 44 sequences of consecutive numbers on a Lottery ticket, the chance of someone getting any consecutive sequence is one in 317,814.

Geoffrey Grimmett, professor of mathematical statistics at Cambridge University, said: "The chance of getting those numbers is the same as any other sequence, but the thing about this game is that there are lots of sequences which are surprises."

This chapter is concerned with probability and extends the work covered in *Intermediate 1*. Here is a reminder of some facts about probability.

- **Theoretical probability** is a way of predicting what we can expect to happen. For example, if a coin is spun 100 times we would in theory expect it to land Heads 50 times and Tails 50 times, as the probability of getting Heads or Tails is equal.
- **Experimental probability** shows the results of an experiment. For example, a coin spun 100 times may be recorded as showing 56 Heads and 44 Tails.
- The probability of an event $= \dfrac{\text{the number of favourable outcomes}}{\text{the total number of equally likely outcomes}}$

  For example, the probability of throwing a 3, P(3), with an ordinary dice is $\frac{1}{6}$. So P(3) $= \frac{1}{6}$.

- If there are n equally likely outcomes, then the probability of each one happening is $\dfrac{1}{n}$ and the probability of any one *not* happening is $1 - \dfrac{1}{n}$.

Exercise 15.1

(*Revision*)

1 500 tickets are sold for a raffle and one winning ticket is drawn. What is the probability of winning if you buy:
   a 1 ticket
   b 5 tickets
   c 500 tickets
   d 0 tickets?

2 In a class there are 16 girls and 14 boys. The teacher takes in all of their books in a random order. Calculate the probability that the teacher will:
   a mark a book belonging to a girl first
   b mark a book belonging to a boy first.

3 Twenty-six tiles, each lettered with one different letter of the alphabet, are put into a bag. If one tile is drawn out at random, calculate the probability that it is:
   a an A or Z
   b a vowel
   c a consonant
   d an A, B, C or D
   e a letter in your surname.

4 **a** Four red, 8 white, 6 blue and 2 green counters are put into a bag. If one is picked at random, calculate the probability that it is:
      i) a green counter   ii) a red counter.
   b If the first counter taken out is green and it is not put back into the bag, calculate the probability that the second counter picked is:
      i) a green counter   ii) a red counter.

5 A roulette wheel has the numbers 0 to 36 equally spaced around its edge. Assuming that it is unbiased, calculate the probability, on spinning it, of getting:
   a the number 7
   b a prime number
   c zero
   d a number greater than 10
   e a multiple of 3 or 5.

Combined events

When we look at **combined events**, we are looking at problems involving two or more events. The outcomes of these events are commonly shown either in a list form or in a **two-way table**.

Example

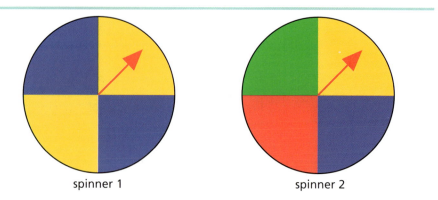

spinner 1   spinner 2

a Two spinners (shown above) are spun. List all the possible outcomes.
b Represent all the possible outcomes in a two-way table.
c What is the probability of getting two blues when the spinners are spun?

a Blue   Blue
       Blue   Red
       Blue   Green
       Blue   Yellow
       Yellow   Blue
       Yellow   Red
       Yellow   Green
       Yellow   Yellow

b

		spinner 2			
		blue	**red**	**green**	**yellow**
spinner 1	**yellow**	yellow, blue	yellow, red	yellow, green	yellow, yellow
	blue	blue, blue	blue, red	blue, green	blue, yellow

c As the outcomes are all equally likely, P(blue, blue) $= \frac{1}{8}$.

Note. In this example the events are **independent** as the result of one spinner doesn't affect the result of the other.

Exercise 15.2

(Revision)

1 a Two spinners are each coloured half black and half white. Write a list describing all the possible outcomes when they are spun.
b Draw a two-way table to show all the possible outcomes of spinning the two spinners.
c Using your results, calculate the theoretical probability of getting a white and a black.

2 a Two fair tetrahedral dice are thrown. If each is numbered 1–4, write a list of all the possible outcomes when rolling the two dice.

b Represent all the possible outcomes in a two-way table.

c What is the theoretical probability of both dice showing even numbers?

d What is the theoretical probability that the number on one dice is 2 more than the number shown on the other dice?

e What is the theoretical probability that the sum of both numbers is an odd number?

3 The letters A, P and T can be combined in several different ways.

a Write all three letters in as many different combinations as possible.

b If a computer writes these three letters at random, calculate the probability that

   i) the letters are written in alphabetical order

   ii) the letter A is written first

   iii) the letter P is written after the letter A

   iv) the computer writes the word 'TAP'.

4 Is the rolling of two dice an example of independent events? Explain your answer.

Tree diagrams

> Tree diagrams are sometimes called probability trees.

Tree diagrams are another good way of representing information diagrammatically, particularly if the events are not equally likely, or if they are **mutually exclusive**.

● *Two outcomes are said to be mutually exclusive if they cannot both happen together.*

Example

> If you spin a coin once you cannot get a Head and a Tail! It's one or the other: they are mutually exclusive.

a If a coin is spun twice, show all the possible outcomes on a tree diagram, writing each of the probabilities along the branches.

b What is the probability of getting two Heads?

c What is the probability of getting a Head and a Tail in any order?

d What is the probability of getting at least one Head?

e What is the probability of getting no Heads?

a

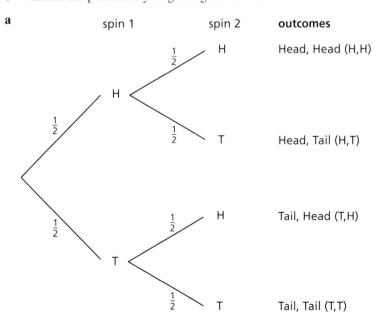

Remember:
$P(H, H) = P(H) \times P(H)$

b There are four equally likely outcomes, therefore the probability of getting $(H, H) = \frac{1}{4}$.

or

$P(H, H) = \frac{1}{2} \times \frac{1}{2} = \frac{1}{4}$.

c The successful outcomes are (H, T) and (T, H). As each of these individually has a probability of $\frac{1}{4}$, the probability of both $= \frac{2}{4}$ or $\frac{1}{2}$.

d This refers to any outcome with either one or two Heads, i.e. all of the outcomes *except* TT.

Therefore the probability $= \frac{3}{4}$.

Remember:
These are the theoretical probabilities. If you tried the experiment four times it is unlikely you would get HH, HT, TH, TT.

Note. This is the same as $1 - \frac{1}{4}$, where the 1 represents all outcomes and $\frac{1}{4}$ represents the probability of getting TT.

e The only successful outcome for this event is TT.

Therefore the probability $= \frac{1}{4}$.

Exercise 15.3

Exercises 15.3 and 15.4 refer to tree diagrams but could also be answered using two-way tables.

1 a A family has two children. Draw a tree diagram to show all the possible combinations of boys and girls.
   b Assuming that the family is equally likely to get a boy as a girl, calculate the probability of getting:
   i) two girls   ii) one girl and one boy
   iii) at least one girl   iv) two boys.

2 a A computer uses combinations of the numbers 1, 2 and 3 to make a random two-digit number. Assuming that a number can be repeated, show on a tree diagram all the possible combinations that the computer can print.
   b Calculate the probability of getting:
   i) the number 13   ii) an even number
   iii) a multiple of 11   iv) a multiple of 3
   v) a multiple of 2 or 3

3 a A netball team plays two matches. In each match it is equally likely to either win, lose or draw. Draw a tree diagram to show all the possible outcomes over the two matches.
   b Calculate the probability that the team:
   i) wins both matches   ii) wins more times than it loses
   iii) loses at least one match   iv) either draws or loses both matches.
   c Explain why in this case it is not a good assumption that all outcomes are equally likely.
   d Are the results of a netball match mutually exclusive? Explain your answer.

4 A spinner is split into quarters.
   a If it is spun twice, draw a probability tree showing all the possible outcomes.
   b Calculate the probability of getting:
   i) two dark blues
   ii) two blues of either shade
   iii) a black and a white in any order.

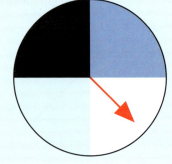

In each of the cases considered so far, all of the outcomes have been equally likely. However, this need not be the case.

Example

a In winter the probability that it rains on any one day is $\frac{5}{7}$.

   i) By drawing a tree diagram, show all the possible combinations for two consecutive days, and find their probabilities.

   ii) Write each of the probabilities along the branches.

b Calculate the probability that it will:

   i) rain on both days

   ii) rain on the first but not the second day

   iii) rain on at least one day.

a

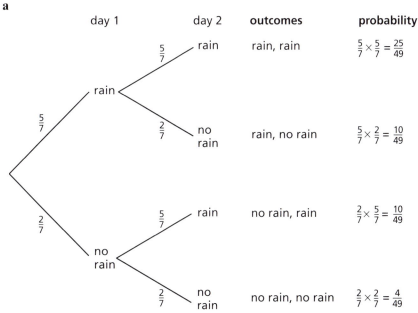

day 1	day 2	outcomes	probability
	rain	rain, rain	$\frac{5}{7} \times \frac{5}{7} = \frac{25}{49}$
rain	no rain	rain, no rain	$\frac{5}{7} \times \frac{2}{7} = \frac{10}{49}$
no rain	rain	no rain, rain	$\frac{2}{7} \times \frac{5}{7} = \frac{10}{49}$
	no rain	no rain, no rain	$\frac{2}{7} \times \frac{2}{7} = \frac{4}{49}$

Note that the probability of each outcome is arrived at by multiplying the probabilities of the branches leading to that outcome.

b i) Rain on both days

     $P(R, R) = \frac{5}{7} \times \frac{5}{7} = \frac{25}{49}.$

  ii) Rain on the first day but not the second day

     $P(R, NR) = \frac{5}{7} \times \frac{2}{7} = \frac{10}{49}.$

  iii) Rain on at least one day

     The outcomes which satisfy this event are (R, R) (R, NR) and (NR, R).

     Therefore the probability is $\frac{25}{49} + \frac{10}{49} + \frac{10}{49} = \frac{45}{49}.$

Exercise 15.4

1 A particular board game involves players rolling a dice. However, before a player can start, he or she needs to roll a 6.
  a Copy and complete the tree diagram below, showing all the possible combinations for the first two rolls of the dice.

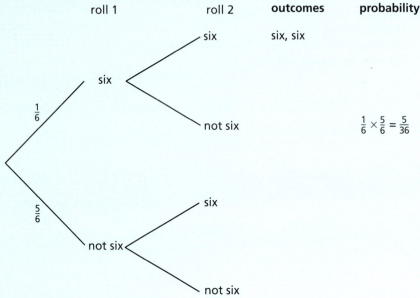

  b Calculate the probability of:
   i) getting a 6 on the first throw
   ii) starting within the first two throws
   iii) starting on the second throw
   iv) not starting within the first two throws.
2 In Italy $\frac{3}{5}$ of cars are foreign made. By drawing a tree diagram and writing the probabilities along each of the branches, calculate the following probabilities.
  a The next two cars to pass a particular spot are both Italian made.
  b The next two cars are foreign.
  c At least one of the next two cars is Italian made.
3 A radio station plays 70% pop, 15% jazz and 15% classical music.
  a Assuming that music is played in a random order, draw a probability tree showing all the possible combinations for the next two pieces of music.
  b Write the probabilities along each of the branches.
  c Calculate the probability that:
   i) the next two pieces played are pop
   ii) one of the next two pieces is jazz
   iii) the next two pieces are both classical
   iv) two different types of music are played.
4 The probability that a morning bus arrives on time is 65%.
  a Draw a tree diagram showing all the possible outcomes for two consecutive mornings.
  b Label your tree diagram with all the relevant information.
  c Use your diagram to calculate the probability that:
   i) the bus is on time on both mornings
   ii) the bus is late on both mornings
   iii) the bus is on time at least once.

5 A normal pack of 52 cards is shuffled and two cards picked at random. Each time a card is picked, its value is noted down and then it is put back in the pack. Draw a tree diagram to help calculate the probability of picking:

  a two Clubs
  b no Clubs
  c at least one Club.

6 A bowl of fruit contains one apple, one banana, two oranges and two pears. Two pieces of fruit are chosen at random and eaten.

  a Draw a probability tree showing all the possible combinations of the two pieces of fruit.
  b Use your tree diagram to calculate the probability that:
    i) both the pieces of fruit eaten are oranges
    ii) an apple and a banana are eaten
    iii) at least one pear is eaten.

7 Light bulbs are packaged in packs of two. 10% of the bulbs are found to be faulty. Calculate the probability of finding two faulty bulbs in a single pack.

8 A volleyball team has a 0.25 chance of losing a game. Calculate the probability of the team achieving:

  a two consecutive wins
  b three consecutive wins
  c ten consecutive wins.

SUMMARY

By the end of this chapter you should know:

- that **theoretical probability** is a way of predicting what we can expect to happen
- that **experimental probability** shows the results of an experiment
- how a **two-way table** or a list can be used to show the outcomes of **combined events**, for example to show all the possible combinations when two ordinary coins are spun

		coin 1	
		H	T
coin 2	H	H, H	H, T
	T	T, H	T, T

- that **independent events** do not affect each other; for example the outcome of one roll of a dice will not affect the outcome of a second roll of the dice
- that **mutually exclusive** events cannot happen together; for example drawing a Diamond or drawing a Club from a pack of cards are mutually exclusive events, but drawing a Diamond or a King from a pack of cards are *not* mutually exclusive events as it is possible to draw the King of Diamonds

■ how **tree diagrams** can be used to show combined events and to calculate the probability of particular outcomes; for example the following tree diagram shows all the possible outcomes of spinning a coin twice and, using the tree diagram, the probability of getting two Heads can be calculated.

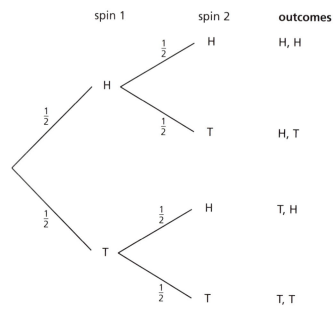

spin 1   spin 2   outcomes

$$P(H, H) = \tfrac{1}{2} \times \tfrac{1}{2} = \tfrac{1}{4}$$

Exercise 15A

1 Calculate the theoretical probability of:
  a being born on a Wednesday
  b being born on the 5th of a month in a non-leap year
  c being born on 20 June in a non-leap year
  d being born on 29 February (in a leap year).

2 If an ordinary, fair dice is thrown, calculate the theoretical probability of getting:
  a a 2
  b an even number
  c a 3 or more
  d less than 1.

3 A bag contains 12 white counters, 7 black counters and 1 red counter.
  a If, when a counter is taken out, it is not replaced, calculate the probability that:
    i) the first counter removed is white
    ii) the second counter removed is red, given that the first was black.
  b If when a counter is picked it is then put back in the bag, how many attempts will be needed before it is mathematically certain that a red counter will be picked out?

4 A coin is spun and an ordinary, fair dice is rolled.
  a Draw a two-way table showing all the possible combinations.
  b Calculate the probability of getting:
    i) a Head and a 6
    ii) a Tail and an odd number
    iii) a Head and a prime number.

5 Two spinners A and B are split into quarters and coloured as shown. Both spinners are spun.

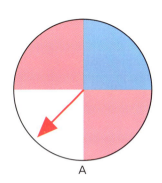
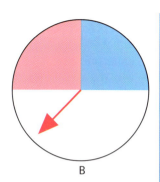

A   B

  a Draw a fully labelled two-way table or tree diagram showing all the possible combinations on the two spinners. Mark the probability of each outcome along the branch.

  b Use your table or diagram to calculate the probability of getting:
    i) two blues   ii) two reds
    iii) a red on spinner A and a white on spinner B.

Exercise 15B

1 If a card is picked from a complete pack of 52 playing cards, calculate the probability of getting:
  a a 9   **b** a Heart
  c the 7 of Clubs   **d** a black Jack, Queen or King.

2 Two normal and fair dice are rolled and their scores added together.
  a Using a two-way table, show all the possible scores that can be achieved.
  b Using your two-way table, calculate the probability of getting:
    i) a score of 12   ii) a score of 7
    iii) a score less than 4   iv) a score of 7 or more.
  c Two dice are rolled 180 times. In theory, how many times would you expect to get a score of 6?

3 A spinner is numbered as shown.
  a If it is spun once, what is the probability of getting:
    i) a 1
    ii) a 2.
  b If it is spun twice, what is the probability of getting:
    i) a 2 followed by a 4
    ii) a 2 and a 4 in any order
    iii) at least one 1
    iv) at least one 2.

4 Two spinners are coloured as shown.

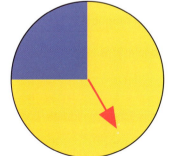
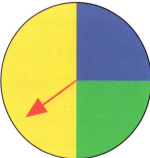

  a They are both spun. Draw and label a two-way table or tree diagram showing all the possible outcomes.
  b Use your table or diagram to calculate the probability of getting:
    i) two blues   ii) two yellows
    iii) a yellow and a green   iv) at least one yellow.

5 A tetrahedral dice numbered 1–4 is rolled twice.
  a Draw a two-way table or tree diagram to show all the possible outcomes.
  b Use your table or diagram to calculate the probability of getting:
    i) two 1s   ii) two numbers the same
    iii) two even numbers.

Exercise 15C

To enter the National Lottery, contestants must choose six different numbers from the numbers 1–49. To win, the six numbers that the contestant chooses must be the six numbers that are drawn at random by a machine.

The calculation for working out the probability of winning the National Lottery jackpot is:

$$\frac{6}{49} \times \frac{5}{48} \times \frac{4}{47} \times \frac{3}{46} \times \frac{2}{45} \times \frac{1}{44} = \frac{1}{13\,983\,816}$$

Show clearly, using a tree diagram, why this calculation is correct.

> That's nearly 14 million to 1 against winning the jackpot!

Exercise 15D

The numbers in the National Lottery are chosen randomly.

You will need:
- computer with internet access
- spreadsheet package

- Visit the National Lottery website at http://www.national-lottery.co.uk/game/results.html
- Enter the 'History of Results' section. Here it is possible to look at all the past results.
- Copy and paste the last six months' results into a spreadsheet. The top of your spreadsheet will look similar to the one shown below:

- Using your spreadsheet, test the following hypotheses:
  - the number 13 is unlucky
  - even numbers are more likely than odd numbers
  - some machines are biased.
- Explain clearly how you reached your conclusions.

Exercise 15E

You will need:
- computer with internet access

Using the internet as a resource, find out how Pascal's triangle is connected with probability theory.

> The study of probability took off in the seventeenth century after an exchange of letters between two French mathematicians, Pierre Fermat and Blaise Pascal.

Review section

Introduction

We have now covered all of the topics required for your final GCSE examination. The problem that we all have at the end of a long course which leads to a final examination is that, although we did the work, understood and could apply the concepts and did reasonably well in tests, much of it has slipped away from our memory.

This review section is intended to help with your revision and to highlight areas of work where you may need to go back to the original chapter or to the A and B exercises and supplementary exercises.

The first part of the review section consists of **20 short-question reviews** grouped into topics as follows:

- 1–8 cover mainly **number**
- 9–12 cover mainly **algebra**
- 13–17 cover mainly **shape**, **space and measures**
- 18–20 cover mainly **probability and handling data**.

 Throughout these reviews, there is an indication of where a calculator should *not* be used.

It is probably not the most efficient, or interesting, way to do the short reviews in numerical order. You may want to do a couple of number reviews, followed by an algebra review, followed by a shape review, and so on. If you come across questions that you cannot do, ask your teacher or refer back to the original chapter. Use the index or contents list to find the topic. Your own exercise books or files should provide you with your own 'worked examples' (that's why your teacher always told you to show your working!).

The final part of the review section consists of **six general reviews of longer (mixed) questions**, similar to the questions you will get in an examination. (Many of these questions should also be done without a calculator, as indicated.) You will probably also have seen sample papers and GCSE papers from previous years.

Revision can be both difficult and boring. The first 20 reviews are deliberately short and should take about one lesson. Research has shown that short periods of revision similar to the reviews included here are the best way of reinforcing learning. So 20 hours or so of calm concentration on these reviews and on sample papers should help to bring your just reward for years of mathematical study. Good luck!

Number

Review 1

1 List the prime factors of 120 in index form.
2 Draw a square side 3.2 units. Use it to show how to calculate $(3.2)^2$.
3 London has a noon temperature of 3 °C on 1 January. Boston is 15 °C colder. What is the temperature in Boston?
4 Change the following fractions to decimals.
    a $\frac{3}{5}$ **b** $\frac{3}{4}$ **c** $\frac{3}{8}$
5 Write the following lengths in order of magnitude, starting with the smallest.

       236 mm   18 cm   0.05 m $\frac{1}{4}$ m

  (Remember: magnitude = size.)
6 Write 85 million in standard form.
7 Copy these equivalent fractions and fill in the blanks.

  a $\dfrac{15}{35} = \dfrac{}{7}$ **b** $\dfrac{3}{} = \dfrac{9}{21} = \dfrac{}{63}$

8 1 mile is 1760 yards. Calculate the number of yards in 15 miles.
9 Calculate the perimeter of a regular hexagon of side 5.2 cm.
10 Find the area and perimeter of a square of side 1.1 cm.

Review 2

You will need:
• graph paper
• ruler

1 The following numbers are expressed correct to two significant figures. Representing each number by the letter x, express the range in which it must lie, by using an inequality:
  a 180   **b** 0.75
2 A machine prints seven books in 50 minutes. How many books will be printed in 5 hours?
3 Find 35% of 560.
4 Change the following times to those shown on a 24-hour clock.
  a A quarter to eight in the morning   **b** 8.30 p.m.
5 In 1999 £1 sterling could be exchanged for 9.5 South African rand. Draw a conversion graph and use it to convert the following.
  a £6 into rand   **b** 60 rand into pounds
6 Express the following in metres.
  a 160 cm   **b** 2.8 km   **c** 12 cm
7 Calculate the simple interest on £650 for three years at 7.5%.
8 Calculate the perimeter of an equilateral triangle of side 13.5 cm.
9 Calculate the volume of a shoebox which is a cuboid with edges 25 cm, 12 cm and 18 cm.

Review 3

You will need:
• graph paper
• ruler

1 What is the highest common factor of 72, 144 and 108?
2 Draw a table of $y = x^2$ for values of x from 0 to +8. Use your table to draw a graph of $y = x^2$. Use the graph to estimate (to 1 d.p.):
  a $(3.2)^2$ **b** $(7.5)^2$ **c** $\sqrt{30}$ **d** $\sqrt{60}$
3 The temperature in Moscow on New Year's Day is -28 °C. The temperature in Sydney on New Year's Day is 28 °C. What is the difference in temperature between the two cities?

4 Change the following decimals to fractions. Give each fraction in its simplest form.
   a 0.25   **b** 0.625   **c** 1.05

5 Represent each of the following inequalities on a number line, where x is a real number.
   a $x<3$ **b** $x\geqslant 4$
   c $1<x<4$ **d** $3>x>-1$

> **Remember:**
> $<$ *less than*
> $>$ *greater than*
> $\geqslant$ *greater than or equal to*

6 Write 0.000 007 83 in standard form.
7 Evaluate $3\frac{2}{5}-1\frac{1}{2}$.
8 Round the following numbers to the number of significant figures shown in brackets:
   a 68.3 (1 s.f.)   **b** 478 700 (2 s.f.)   **c** 645.380 (1 s.f.)   **d** 645.380 (3 s.f.)
9 Calculate the perimeter of a regular octagon of side 17.5 cm.
10 Calculate the area and perimeter of a square of side 7.5 cm.

Review 4

You will need:
- graph paper
- ruler

1 Find the next two terms in each of the following sequences.
   a 108, 96, 84, 72, …, …
   b 1, 2, 3, 5, 8, 13, 21, …, …
2 The following numbers are expressed to the nearest whole number. Illustrate on a number line the range in which each must lie.
   a 6   **b** -4 **c** 100
3 A bricklayer lays 1200 bricks in a 9-hour day. Assuming he works at the same rate all day, how many bricks will he lay in 45 working hours?
4 Increase 450 euro by 8%.
5 A bus journey from Cambridge to Bourn takes 1 hour and 45 minutes. At what time will a bus leaving Cambridge at 08.50 arrive in Bourn?
6 At the beginning of 2002, 1 US dollar could be exchanged for 1.7 Swiss francs. Draw a conversion graph and use it to convert the following.
   a 12 dollars into francs   **b** 150 francs into dollars
7 Express the following masses in kg.
   a 28 000 g   **b** 2400 g   **c** 150 g
8 How long will it take a sum of 5000 euro invested at 4% to earn simple interest of 800 euro?
9 Calculate the perimeter of a regular pentagon of side 3.8 cm.
10 Calculate the volume of a cube of side 1.5 cm.

Review 5

1 What is the lowest common multiple of 7, 8 and 12?
2 Calculate the following square roots without using a calculator.
   a $\sqrt{49}$ **b** $\sqrt{0.81}$ **c** $\sqrt{0.04}$
3 A helicopter hovering at 450 m above sea level drops a sonar device onto the ocean floor. If the ocean is 1680 m deep at this point, how far above the sonar device is the helicopter?
4 Work out 345×17.
5 A month has at least 28 days, but not more than 31 days. Illustrate this information using inequalities.
6 Write the answer to the calculation of 200 000 multiplied by 6 500 000 in standard form.
7 Evaluate $6\frac{3}{7}\div 1\frac{1}{14}$.
8 Estimate the answer to $\dfrac{3.8\times 25.9}{1.6\times 6.8}$.
9 Calculate the perimeter of a decagon of side 4.3 cm.
10 Calculate the volume of a cube of side 1.1 cm.

Review 6

You will need:
• graph paper
• ruler

1 A girl's mass was measured to the nearest 0.1 kg. If her mass was 48.2 kg, illustrate on a number line the range within which it must lie.
2 A paint mix uses red and white paint in a ratio of 3 : 10. How much red paint should be mixed with 7.5 litres of white paint?
3 A television set priced at £280 is reduced by 30% in a sale. What is the new price?
4 Express 'a quarter to four in the afternoon' as it would appear on a 24-hour clock.
5 At the beginning of 2002, 1 US dollar could be exchanged for 7.75 Hong Hong dollars. Draw a conversion graph and use it to convert the following.
   a 12 US dollars into Hong Kong dollars   **b** 50 Hong Kong dollars into US dollars
6 Express the following in litres.
   a 8500 ml   **b** 400 ml   **c** 15 500 ml
7 What rate of simple interest per year must be paid on a principal of 375 euro, in order to earn 90 euro in four years?
8 Calculate the area and perimeter of a square with side of length 6.5 cm.
9 Calculate the volume of a cuboid 12 cm long, 8 cm wide and 8 cm deep.
10 Calculate the total surface area of a cuboid 12 cm long, 8 cm wide and 8 cm deep.

Review 7

1 Evaluate the following.
   a $(-6)+(-4)$ **b** $(-6)\times(-4)$ **c** $(-54)\div(+9)$
2 Without using a calculator, find the value of:
   a $\sqrt{\frac{36}{49}}$ **b** $\sqrt{1\frac{19}{81}}$ **c** $\sqrt{6\frac{1}{4}}$
3 The library of Celsius in Ephesus was built in 380 BC. How old is this building?
4 Work out $4840\div23$, giving your answer to 1 d.p.
5 Write the following fractions in order of size, starting with the smallest.

   $\frac{1}{2}$ $\frac{2}{3}$ $\frac{4}{7}$ $\frac{3}{5}$ $\frac{5}{9}$

6 The light from a star takes four years to reach Earth. If the speed of light is 3×10^5 km/s, calculate the distance of the star from Earth. Give your answer in kilometres in standard form.
7 Evaluate $1\frac{1}{2}\times3\frac{3}{4}\div\frac{7}{8}$.
8 A rectangle's dimensions are given as 8.3 m by 58.2 m. *Estimate* its area.
9 Calculate the perimeter of a regular heptagon (seven sides) of side 6.3 cm.
10 Calculate the volume of a cube of side 2.4 cm.

Review 8

You will need:
• graph paper
• ruler

1 For each of the sequences shown below give an expression for the *n*th term.
    a 7, 13, 19, 25, 31, ...
    b 0, 5, 10, 15, 20, ...

2 The numbers below are rounded to the degree of accuracy shown in brackets. Express the lower and upper bounds of each of these numbers as an inequality.
    a $x = 4.55$ (2 d.p.)   **b** $y = 10.0$ (1 d.p.)

3 Divide 28 m in the ratio 3:4.

4 Find 75% of 640.

5 How long will it take a train travelling at an average speed of 120 km/h to travel 300 km?

6 At the beginning of 2002, £1 sterling could be converted into 2.7 Australian dollars. Draw a conversion graph and use it to convert the following.
    a £25 into Australian dollars
    b 50 Australian dollars into pounds sterling

7 Express the following lengths in metres.
    a 650 cm   **b** 0.85 cm   **c** 15 cm

8 Calculate the simple interest on £340 for four years at 3.75%.

9 Calculate the perimeter and area of a rectangular field 110 m by 75 m.

10 Calculate the volume of a cube of side 3.5 cm.

Algebra

Review 9

You will need:
• graph paper
• ruler

1 Find the value of the following without using a calculator.
   a 2^3 **b** 3^4 **c** $2^4 \times 3^2$

2 Expand and simplify:
   a $3(2m - 6) + 2(3 - 4m)$ **b** $2x(x - y) - 2y(y - x)$

3 Factorise:
   a $15ab - 10bc$ **b** $6p^2q + 9pq^2 + 12pq$

4 If $x = 2$, $y = 3$ and $z = 5$, evaluate:
   a $3x - 4y + 2z$ **b** $x^2 - y^2 - z^2$

5 Rearrange the following formulae to make the letter in brackets the subject.

   a $3c = 2b + a$ (a) **b** $\dfrac{2b + c}{3} = 4a$ (b)

6 Solve the following equations.

   a $3a + 4 = 19$ **b** $\dfrac{b + 4}{3} = 1$

7 Solve the following pair of simultaneous equations.

       $a + 3b = 11$
       $3a - 3b = -3$

8 Given that 50 miles ≈ 80 km, draw a conversion graph up to 100 miles. Use your graph to estimate:
   a how many kilometres there are in 70 miles
   b how many miles there are in 70 km.

9 Sketch the graphs of:
   a $x = 2$ **b** $y = -3$

10 Sketch the graph of $y = x^2$ from -3 to $+3$. Use your graph to find $\sqrt{6.25}$.

Review 10

You will need:
• graph paper
• ruler

1 Find the value of the following without using a calculator.

   a 6^2 **b** 5^3 **c** $\dfrac{2^5}{4^2}$

2 Expand the following and simplify if possible.
   a $4(2a - 3b) + 5(2b - 3a)$ **b** $a(2b - 4) + b(5 - 2a)$

3 Factorise:
   a $12x^2y - 48xy^2$ **b** $3abc - 6b^2c + 9bc^2$

4 If $p = 2$, $c = -3$ and $d = -1$, evaluate:
   a $2p - c + 3d$ **b** $p^2 + c^2 - d^2$

5 Rearrange the following formulae to make the letter in brackets the subject.

   a $2p = 3q - r$ (r) **b** $x - y = \dfrac{x - z}{3}$ (z)

6 Solve the following equations.

a $4a - 9 = -1$

b $\dfrac{b - 5}{2} = 2$

7 Solve the following simultaneous equations.

$$2a - 2b = 4$$
$$4a - b = 5$$

8 Given that $0\,°C = 32\,°F$ and $100\,°C = 212\,°F$, draw a conversion graph between degrees Celsius and degrees Fahrenheit. Use your graph to convert:
a $50\,°C$ to Fahrenheit
b $100\,°F$ to Celsius

9 Sketch the graphs of:
a $y = 2x$
b $y = -x$

10 Sketch the graph of $y = -x^2$ from -3 to $+3$. Use your graph to find $-\sqrt{5.29}$.

Review 11

You will need:
- graph paper
- ruler

1 Work out the value of the following without using a calculator.
a 7^2
b 2^6
c $3^3 \times 2^3$

2 Expand the following and simplify if possible.
a $5x(x - y + z)$
b $3(a + 2b) - a(2b + 4)$

3 Factorise:
a $39xy - 52yz$
b $4pqr^2 + 8pq^2r - 16p^2qr$

4 If $a = 3$, $b = -3$ and $c = 6$, evaluate:
a $a^3 + b^3$
b $a^4 + abc - c^2$

5 Rearrange the following formulae to make the letter in brackets the subject.

a $3b - c = 2d - e$ (d)

b $\dfrac{p}{q} = rs$ (q)

6 Solve the following equations.

a $8 = 6a - 1$

b $3 = \dfrac{2c - 1}{4}$

7 Solve the following simultaneous equations.

$$a + b = -1$$
$$a - b = 5$$

8 A maths examination is marked out of 140. Draw a conversion graph to convert the following scores out of 140 into percentages.
a 60
b 100
c 125

9 Sketch the graph of $y = \dfrac{1}{x}$ for $-3 \leqslant x \leqslant 3$.

10 a Copy and complete the table below for the function $y = x^2 - 3$.

x	-3	-2	-1	0	1	2	3
y	6			-3			

b Plot the graph of the function $y = x^2 - 3$ for $-3 \leqslant x \leqslant 3$.

Review 12

You will need:
• graph paper
• ruler

1 Work out the value of each of the following without using a calculator.

a 9^2 **b** 10^4 **c** $\dfrac{9^2}{3^4}$

2 Factorise:
  a $17x^2 - 51xy$ **b** $18abc - 27bc^2 + 36b^2c$

3 Expand the following and simplify if possible.
  a $3a(2b - 3) + 3b(3 - 2a)$ **b** $\frac{1}{2}(6a + 4) - \frac{1}{3}(9a - 6)$

4 If $p = -2$, $q = -3$ and $r = -5$, evaluate:
  a $p^2 + q^2 + r^2$ **b** $2pqr - 3qr$

5 Rearrange the following formulae to make the letter in brackets the subject.
  a $4a - 3b = c - 5d$ (d)

  b $\dfrac{m}{2n} = \dfrac{r}{3s}$ (s)

6 Solve the following equations.
  a $3 = 4a - 5$

  b $4 = \dfrac{5b - 3}{8}$

7 Solve the following simultaneous equations.

$$a + b = 9$$
$$2a - 2b = 2$$

8 At the beginning of 2002, £1 sterling would convert to 1.6 euro. Draw an appropriate conversion graph to convert the following:
  a £70 to euro **b** 200 euro to pounds

9 Sketch the graph of $y = -3x$

10 a Copy and complete the table below for the function $y = x^2 + x$.

x	−3	−2	−1	0	1	2	3
y	6				2		

  b Plot the graph of the function $y = x^2 + x$ for $-3 \leqslant x \leqslant 3$.

Shape, space and measures

Review 13

You will need:
• ruler
• protractor
• squared paper

1 Draw an obtuse angle of approximately 120°. Then, using a protractor, measure the angle exactly.

2 Construct an isosceles triangle of base length 5.2 cm and two sides of 8 cm. Measure the base angles.

3 Calculate the size of each exterior angle of a regular pentagon.

4 The diagram below shows a trapezoidal garden. Three of its sides are enclosed by a fence, while the fourth is next to a house. Grass seed is to be sown in the garden. However, the grass must be at least 3 m away from the house and at least 1.5 m away from the fence. Draw the diagram accurately to scale and shade the region in which the seed can be sown.

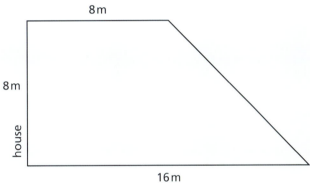

5 Draw the five vowels as capital letters and show all their lines of symmetry.

6 Copy the diagram below and draw in the position of the image when the object is reflected in the dotted line.

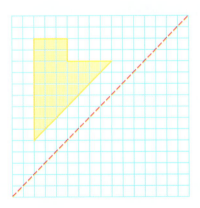

Review 14

You will need:
• protractor
• ruler

1 Calculate the area of the kite below.

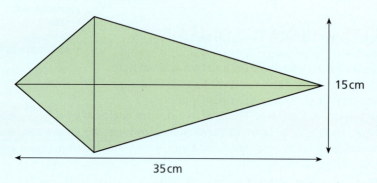

15 cm

35 cm

2 The wheel of a bicycle has a radius of 32 cm. Calculate the distance, in km, the bicycle has travelled after 2000 revolutions. Give your answer to 2 s.f.

3 Use a scale of 1 cm to 1 km to show the following journey of a yacht.
Starting at point A, it sails 8 km on a bearing of 135° to point B.
From B, it sails 10 km on a bearing of 280° to point C. Find the
distance and bearing of A from C.

Remember:
*Bearings are always given
as three figures.*

4 Calculate the size of the angle ABC in the triangle below. Give your answer to 2 s.f.

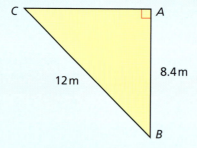

C

A

8.4 m

12 m

B

5 Using the diagram below, describe the following translations using column vectors.

a $\overrightarrow{AB}$ **b** $\overrightarrow{CD}$

c $\overrightarrow{DE}$ **d** $\overrightarrow{EA}$

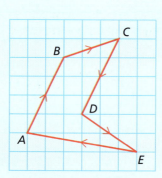

Review 15

You will need:
• ruler
• compasses
• squared paper

1 a What is the supplement of 60°?
  b What is the complement of 60°?
2 a Draw a line AB 8.8 cm long. Construct the perpendicular bisector of AB.
  b Draw a line PQ about 8 cm long. Put a point O roughly one third of the way along the line. Construct a perpendicular to O.
  c Draw another similar line PQ. Put a point O about the same distance along the line but 5 cm above PQ. Construct a perpendicular from O to PQ.

> **Hint:**
> *Put your compasses on O and make arcs cutting PQ.*

3 The interior of a regular polygon is 150°. Name the polygon and sketch it.
4 Draw a circle of radius 5 cm. Inside this circle, label points A, B and C, ensuring that they do not lie in a straight line.
  a By construction, find the point which is equidistant from A, B and C. Label this point O.
  b Mark two points D and E on the circumference of the circle, such that $OD = OE = OA = OB = OC$. If no such points exist in your diagram, explain why not.
5 Draw a two-dimensional shape with rotational symmetry of order 6. Mark on the shape all its lines of symmetry.
6 In the diagram shown, triangle $A'B'C'$ is an enlargement of triangle ABC.
  Copy the diagram and
  a find the position of the centre of enlargement
  b calculate the scale factor of enlargement.

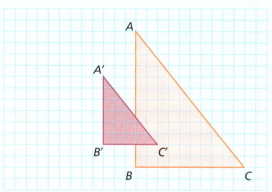

Review 16

You will need:
• squared/graph paper

1 Calculate the height of a trapezium with area 20 cm², if its parallel sides measure 6 cm and 4 cm.
2 Calculate the volume, in cm³, of a cylinder of radius 4 cm and length 12 cm. Give your answer to 3 s.f.
3 Use Pythagoras' theorem to calculate the length of the side marked x cm in the diagram below. Give your answer to 1 d.p.

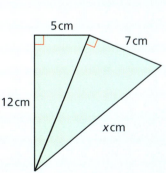

4 Use trigonometry to calculate the size of the angle marked y in the diagram below.

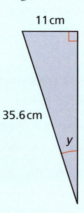

11 cm

35.6 cm

y

5 **p** is represented by the column vector $\begin{pmatrix} 2 \\ 5 \end{pmatrix}$.

q is represented by the column vector $\begin{pmatrix} -4 \\ -2 \end{pmatrix}$.

a Draw a diagram to show the single transformation $\mathbf{p} + \mathbf{q}$.
b What column vector is represented by $\mathbf{p} + \mathbf{q}$?

Review 17

You will need:
• protractor
• ruler

1 Use a protractor and a ruler to draw a triangle ABC with $AB = 8.4$ cm, $\angle BAC = 40°$ and $\angle ABC = 80°$. Measure BC, AC and $\angle BCA$.
2 Identify the two quadrilaterals described below.
  a Opposite sides are equal in length and diagonals bisect each other. Adjacent angles are not equal.
  b All sides are equal in length. Diagonals, however, are not.
3 Draw a scale diagram of a rectangular garden 8 m by 12 m. Identify the locus of points which fulfil all the following criteria:
  a at least 1.5 m from the edge of the garden
  b at least 3.5 m from the corners of the garden
  c at least 2 m from the centre of the garden.
4 Draw a shape made from four congruent, equilateral triangles and one square, which has rotational symmetry of order 2 and only two lines of symmetry.
5 In the diagram below, O marks the centre of the circle. Calculate the size of the angle marked x.

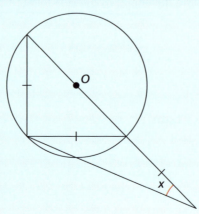

O

x

Probability and handling data

Review 18

You will need:
• graph paper or squared paper

1 Find the mean, median, mode and range of each of the sets of numbers below.

a 1	4	4	5	6	7	8				
b 4	12	7	9	3	4	8	7	16	9	4	10
c 6.4	8.2	7.3	6.3	8.2	8.7	5.6	8.2			

2 An ordinary dice is thrown 120 times. Calculate the number of times you would expect to throw:
  a a 6
  b an even number.

3 On a Saturday in November, the football results for the English Premier League showed that there were eight home wins, six draws and four away wins. Represent this data on a pie chart.

4 The table below shows the marks out of 100 in a French test for a class of students. Draw a bar chart representing this data.

score	frequency
1–20	3
21–40	4
41–60	6
61–80	11
81–100	7

5 The numbers below are the number of words in the first 40 sentences of Macbeth:

5	6	10	8	3	3	5	3	2	13
5	3	11	15	3	12	15	30	32	26
18	16	22	14	11	7	9	11	15	17
21	28	17	15	9	7	5	23	27	34

Record this data on a frequency table, using intervals 1–10, 11–20, etc. Draw a frequency diagram for the data.

6 A roulette wheel has the numbers 1 to 36 and zero. When the wheel is spun, what is the probability of the ball landing on:
  a a four
  b an even number
  c zero
  d a factor of 36
  e a prime number.

See over for questions **7–10**.

7 A motor car manufacturer publishes a pie chart. It represents their customers' choice of colour for their new small car in one year. If the number of new cars in the survey was 72 000, how many cars were:
   a red   **b** white?

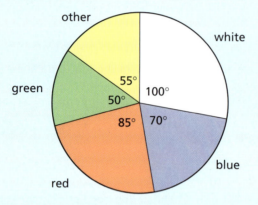

8 Illustrate the data in question 7 in a suitable form other than a pie chart.
9 Seats for a school concert are arranged in rows labelled A to K, with 15 seats in each row numbered 1–15. Every seat is occupied, and every person is given a ticket for a raffle. Calculate the probability that the first prize is won by:
   a seat J4
   b a seat in row B
   c a person sitting in the middle of a row.
10 Explain the difference between a 'fair' game and an 'unfair' game.

Review 19

> **You will need:**
> ● graph paper or squared paper

1 Find the mean shoe size of a group of students whose shoe sizes are recorded below.

size of shoe	4	5	6	7	8	9	10
number of students	3	4	8	6	8	4	1

2 The number of letters received on one day by people living in one street is recorded below. Display this information on a suitable pictogram.

letters received	0	1	2	3	4	5
number of houses	6	4	2	3	2	1

3 Display the information in question 2 on a pie chart.
4 A school orchestra has the following number of players in each section.

type of instrument	strings	woodwind	brass	percussion
number of players	12	10	6	2

Illustrate this data on a bar chart.
5 Illustrate the data in question 4 in another suitable way.
6 A dice in the shape of an icosahedron (20 faces) is thrown 100 times. The faces are numbered 1 to 20. How many times would you expect the top face to show:
   a a 12   **b** an odd number   **c** a factor of 20?

7 Lettered tiles, which spell the word AUSTRALIA, are put into a bag. One tile is removed. Calculate the probability that it is:

   a an A   **b** a vowel   **c** a consonant

8 15000 tickets are sold for a raffle to win a car. How many tickets must I buy to have a 5% chance of winning the car?

9 Explain the following terms:

   a range   **b** mode

   c grouped frequency   **d** pictogram

10 A village council is campaigning to have a new secondary school built in the village. Suggest four pieces of useful data that could be collected from people in the area by means of a questionnaire.

Review 20

You will need:
• graph paper or squared paper

1 Find the mean, median and mode of each of the following sets of data.

   a 6   7   8   9   10

   b 3.5   4.5   4.5   5   5.5   6   6.5

   c 8   32   16   8   24   16   8   40   32   8   24

2 A school library recorded students' choices of books under five headings as shown below. Display this data on a suitable pictogram.

crime	28
sci-fi	45
romance	32
other fiction	20
non-fiction	55

3 Display the data in question 2 on a pie chart.

4 Eighty students were asked to record the number of minutes they spent listening to music in one week. The results are shown below.

6	18	32	21	58	47	12	28	56	33	41	15	47	52
32	26	45	59	52	36	43	25	17	19	46	35	58	20
49	30	42	9	19	24	39	51	42	14	17	34	50	59
7	38	16	47	38	8	55	48	36	23	12	28	35	40
59	5	34	13	10	57	45	26	38	46	23	16	51	28
47	32	57	11	9	19	30	42	22	29				

Draw a grouped frequency table with intervals $0 < x \leqslant 10$; $10 < x \leqslant 20$, etc., then draw a bar chart to illustrate the data.

5 540 people were surveyed for hair colour. This information was displayed on a pie chart. If the size of the sector for black hair was 200° and for blond hair 44°, how many people had:

   a black hair   **b** blond hair?

6 A fair octahedral dice is thrown. The faces are numbered 1 to 8. What is the probability of throwing:

   a a 6   **b** a prime number   **c** a number that is not prime?

7 Lettered tiles, which spell out the word MISSISSIPPI, are put into a bag. If one tile is removed, what is the probability that it is:

   a an S   **b** an I   **c** not a vowel?

8 If I buy ten tickets for a raffle, what is the probability that I will win first prize if 400 tickets have been sold?

9 Describe briefly how you would conduct a survey to test the truth of the statement 'all boys like football'.

10 Explain briefly how you might display the data obtained in question 9.

General review 1

1 Construct an equilateral triangle ABC of side 6 cm. Construct further
equilateral triangles on sides AB, BC and CA. The shape produced is a net.

You will need:
- ruler
- compasses
- graph paper

  a Which solid shape is this a net for?

  b Mark the lines of symmetry of the net.

  c What is the order or rotational symmetry of the net?

  d Calculate the surface area of the net.

2 a Plot the graph of $y = x^2$ for values $-4 \leqslant x \leqslant 4$.

  b On the same grid, plot the graph of $y = x + 6$.

  c Use your graph to solve the quadratic equation $x^2 - x - 6 = 0$.

3 The temperature $T°C$ at height H metres above sea level is given by the formula

$$T = 20 - \frac{H}{150}$$

  a Calculate the temperature at 3450 metres.

  b Rearrange the formula to make H the subject.

  c At what height in metres is the temperature $12°C$?

4 A vet has a salary of £57 600 per year. Her tax deductions amount to 30% of her salary, and a further
6% of her salary is deducted for her pension.

  a Calculate

    i) her monthly salary before any deductions are made (gross monthly salary)

    ii) her salary after deductions are made (net monthly salary).

  b Her expenditure for one month is shown below. Illustrate this on a pie chart.

    Tax 30%
    Pension 6%
    Rent £1200
    Food £576
    Clothing and entertainment £960
    The rest is saved.

  c If the vet is awarded a pay rise of 9%, calculate her new gross annual salary.

5 Nineteen pupils take a mathematics test. Their percentages are shown below:

21	40	68	62	94	54	67	74	88	47
55	56	67	52	94	88	36	42	55	

The first four results have been entered onto the stem-and-leaf diagram below.

```
1 |                         key
2 | 1                    2 | 1 means 21
3 |
4 | 0
5 |
6 | 2 8
7 |
8 |
9 |
```

  a Copy and complete the diagram for the remainder of the results.

  b Use your stem-and-leaf diagram to calculate the median test result.

  c Explain how you arrived at your answer to **b** above.

General review 2

1 Golf balls are sold in packs of three either in rectangular boxes or cylindrical tubes as shown

You will need:
• ruler
• compasses

 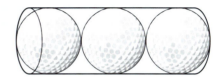

The balls fit their containers exactly. A golf ball has a diameter of 4.2 cm and a volume of 39 cm^3.
a Write down the dimensions of the rectangular box and hence calculate its volume.
b Calculate the percentage volume of the rectangular box occupied by the golf balls.
c Calculate to the nearest cm^3 the volume of the cylindrical container, given the formula $V = \pi r^2 h$, where V cm^3 is the volume of a cylinder of radius r cm and height h cm.
d A tennis ball has a volume of 117 cm^3. Express the volume of a golf ball and the volume of a tennis ball as a ratio in the form $1:n$.

2 The frequency table below gives the scores out of 10 achieved by a year group of students in a maths test.

test score	0	1	2	3	4	5	6	7	8	9	10	total
frequency	3	6	9	6	9	15	12	18	12	3	3	
frequency × score												

a Copy and complete the table.
b Calculate the mean test score.
c Calculate the median test score.
d State the modal test score.

3 Two lines AC and BD bisect each other at right angles. Construct the shape $ABCD$, given that $AC = 12$ cm and $BD = 16$ cm.
a Name the shape $ABCD$.
b What is the order of rotational symmetry of $ABCD$?
c Measure the length of the side AB and confirm this by calculation using Pythagoras' theorem.
d Use trigonometry to calculate angle ABC.

4 a Solve the equation $\dfrac{5 + 2x}{3} = \dfrac{4x - 1}{5}$.

b Solve these simultaneous equations.

$$2x + 3y = 14$$
$$3x - 2y = 19$$

c Rearrange this formula to make a the subject: $\dfrac{a + b}{c} = d - e$.

General review 3

1 The sector below represents the marked area for a discus event in a school sports day.

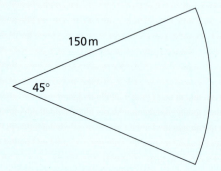

150 m

45°

  a Make a scale drawing of the sector.
  b For safety purposes, a rope is staked outside the sector so that it is always 10 m from the marked sector. Show this on your diagram.
  c The winning throw is 125 m and lands on the line of symmetry of the sector. Label this point *A* on your diagram. Use trigonometry to calculate the shortest distance from *A* to the side line of the sector.

2 a Find the next two terms in each of the following sequences.
    i) 6, 11, 16, 21, ...
    ii) 3, 12, 48, 192, ...
  b The *n*th term of a sequence is given by the formula $(n+1)(n-1)$. Find:
    i) the 5th term
    ii) the 99th term.
  c A sequence of numbers is 1, 8, 27, 64, ...
    i) Write the next two terms.
    ii) Write the 100th term.
    iii) Write the *n*th term.
  d Write the *n*th term of the sequence 6, 13, 32, 69, ...
3 The letters A, D, R, T can be combined as four letters in a number of ways.
  a How many possible combinations of the four letters are there?
  b If the letters are written by a computer at random what is the probability that:
    i) the letters will spell DART
    ii) the letter A will be written first?

4 A rugby team scores the following number of points in 15 matches.

21, 17, 6, 12, 42, 17, 13, 28, 17, 20, 23, 14, 23, 21, 42

a Calculate:
  i) the mean score
  ii) the median score
  iii) the modal score.
b Use the shape of a rugby goal (H) to illustrate the results on a suitable pictogram.

5 Draw the diagram below to scale.

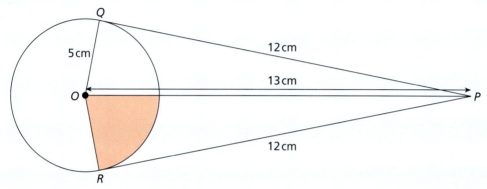

a Prove that *PQ* is a tangent to the circle.
b Use trigonometry to calculate angle *POR*.
c Calculate the area of the circle.

General review 4

1 Copy the graph below.

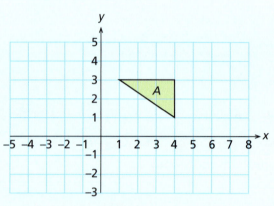

a Reflect triangle *A* in the line *y* = 0. Label it *B*.

b Draw the triangle formed when triangle *A* is rotated through 180° about the origin. Label it *C*.

c Draw the translation of triangle *C* by the vector $\begin{pmatrix} -1 \\ 5 \end{pmatrix}$. Label it *D*.

d Draw the enlargement of triangle *A*, with centre (1, 1) and scale factor 2. Label it *E*.

2

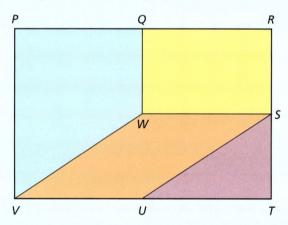

PRTV is a rectangle 100 m by 70 m. *Q* is the midpoint of *PR*. *S* is the midpoint of *RT*. *U* is the midpoint of *TV*. *W* is the midpoint of *QU*.

Calculate:

a the area of *QRSW*

b the area of *WSUV*

c the area of *PQWV*

d the length *VW*

e angle *TUS*.

3 a Expand and simplify the following expression.

$$\frac{a}{9}(18a - 27) - \frac{a}{3}(6a - 12)$$

b Factorise the following: $39x^2y^2z - 26xy^2z^2 + 52x^2yz^2$.
c If $a = 2$, $b = -2$ and $c = 5$ evaluate the expression $a^2b - b^3 - c^2$.
d Rearrange the formula below to make p the subject.

$$3m + n = r(p - q)$$

4 a Construct a regular hexagon of side 5 cm.
b Use your hexagon to illustrate that the sum of the internal angles of a regular hexagon is given by $180(6 - 2)°$.
c How many lines of symmetry does a regular pentagon have?
d What is the order of rotational symmetry of a regular dodecagon?

General review 5

1

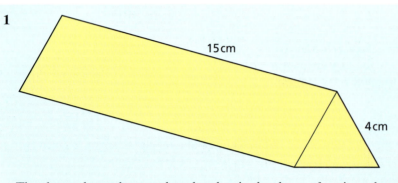

15 cm

4 cm

The shape above shows a chocolate bar in the shape of a triangular prism. Its end face is an equilateral triangle of side length 4 cm. The prism is 15 cm long.

a Draw a scale drawing of the prism's net.

b Calculate the height of the triangular face.

c Calculate the area of the triangular face.

d Calculate the total surface area of the prism.

Six of the chocolate bars fit together to form a hexagonal-faced prism.

e Calculate the surface area of this new prism.

2 A coin is spun twice.

a List all the possible combinations of the two spins.

By experiment, the coin is found to be biased. It comes down Heads, on average, two times out of five.

b Draw a two-way table or tree diagram and use it to calculate the probability of at least one Head occurring from two spins.

3 a Expand and simplify the following expression.

$$\frac{a}{4}(6a - 3) - \frac{a}{8}(16 - 4a)$$

b Rearrange the following formula to make a the subject:

$$v = ut - \tfrac{1}{2}at^2$$

c Factorise $9a^2 - 25b^2$.

4 The duration of telephone calls by two tele-salesmen are recorded as shown below.

<table>
<tr><th colspan="2" align="center">Salesman A</th><th colspan="2" align="center">Salesman B</th></tr>
<tr><th>duration (min)</th><th>frequency</th><th>duration (min)</th><th>frequency</th></tr>
<tr><td>$0 \leqslant t < 5$</td><td>0</td><td>$0 \leqslant t < 5$</td><td>6</td></tr>
<tr><td>$5 \leqslant t < 10$</td><td>2</td><td>$5 \leqslant t < 10$</td><td>6</td></tr>
<tr><td>$10 \leqslant t < 15$</td><td>8</td><td>$10 \leqslant t < 15$</td><td>8</td></tr>
<tr><td>$15 \leqslant t < 20$</td><td>25</td><td>$15 \leqslant t < 20$</td><td>10</td></tr>
<tr><td>$20 \leqslant t < 25$</td><td>9</td><td>$20 \leqslant t < 25$</td><td>8</td></tr>
<tr><td>$25 \leqslant t < 30$</td><td>3</td><td>$25 \leqslant t < 30$</td><td>9</td></tr>
<tr><td>$30 \leqslant t < 35$</td><td>3</td><td>$30 \leqslant t < 35$</td><td>3</td></tr>
</table>

   a Plot a cumulative frequency curve for each salesman.
   b Calculate the median duration of call for each salesman.
   c Use your graphs to calculate the interquartile range for the calls of each salesman.

5

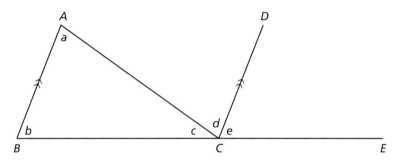

Copy and complete the following sentences based on the diagram above.
a Line AB is parallel to line ...
b $\angle BAC = \angle ACD$ because they are ...
c $\angle ABC = \angle DCE$ because they are ...
d $c + d + e = \dots$ because ...
e $a + b + c = \dots$
f These steps prove that ...

General review 6

1 A company has a contract to build a tunnel under a river 2 km wide. The tunnel will be a cylinder 80 m wide and 2.8 km long. The volume of a cylinder is given by the formula $V = \pi r^2 h$, where r is the radius and h the height (or length) of the cylinder.

You will need:
• graph paper
• ruler
• compasses
• protractor

  a How many cubic metres of earth must be removed in preparation for the tunnel?

  b The earth is removed by trucks which each carry 25 m³ of earth. How many truck journeys are required?

  c The original cost of the tunnel was estimated at £180 million, spread evenly over three years. However, costs increased by 20% in the first year, 15% in the second year and 15% in the third year. Calculate the final cost of the tunnel.

2 A normal pack of playing cards contains 52 cards with 4 suits of 13 cards.

  a If a card is picked at random, what is the probability of picking:

    i) a 7

    ii) a Club

    iii) a picture card

    iv) a 5 or a Heart?

  b If all the picture cards are removed, what is the probability now of picking:

    i) a 7

    ii) a Club

    iii) a picture card

    iv) a 5 or a Heart?

3 ABC is an equilateral triangle of side 13 cm. ACD is a right-angled triangle in which CD is 12 cm.

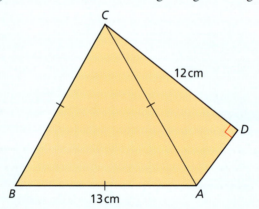

Calculate:

a angle ABC

b the height of triangle ABC

c length AD

d angle ACD.

4 a Sketch the graph of $y = x^2$.

b Copy and complete the table below for the quadratic equation $y = x^2 - 9x + 18$.

x	0	1	2	3	4	5	6	7
y	18		4			−2		

c Plot the graph of $y = x^2 - 9x + 18$ for the range $0 \leqslant x \leqslant 7$.

d Use your graph to solve the following quadratic equation:

$$x^2 - 9x = -18$$

5 A triangle ACB has the following dimensions:

$\angle CAB = 20°$
$AB = 6\,\text{cm}$
$BC = 4\,\text{cm}$

Using an angle measurer, ruler and pair of compasses, construct the two *different* possible triangles.

Index